'The multifunctional land u[se] urgent need to invest in diver[se sus]tainable, productive, climate s[mart...] the research highlights the importance of considering the varying circumstances of vulnerable communities when devising interventions and actions.'

Sithembile Ndema Mwamakamba, Food, Agriculture and Natural Resources Policy Analysis Network (FANRPAN)

'This case-study approach to shifting patterns of cultivation and multifunctional land use lends new insights into food security. From peri-urban agroforestry to watershed approaches to soil conservation, the book demonstrates the potential of both land-owner initiated and state sponsored schemes to simultaneously improve ecosystem services and food provision.'

Professor Andrea Nightingale, University of Oslo, Norway

'In this book a team of young African research colleagues move scientific findings towards policy and practice. They display new ways how to view food security, especially in relation to land use and multifunctional landscapes. AgriFoSe2030 is proud to support this innovative thinking about how to improve Food Security in Africa.'

Professor Ulf Magnusson, Director, AgriFoSe2030

'Tackling multifunctionality in land use, at smallholders' farming context of Africa, is just like hauling back important forgotten policies on sustainable food and nutrition for the poor. Nothing is more important in transformative science than evidence. The set of studies in this book shows facts of dealing with complex landscape aspirations that take us beyond the mere discursive intentions. If you are looking for information about how production at the local scale is influenced by various geographies, social behaviour, marketed drives, and cultural beliefs, get this book as guidance in content and methods to address what most national policies do not often mention in their sectoral approaches.'

Cheikh Mbow, Executive Director of START International

Multifunctional Land Uses in Africa

This book presents contemporary case studies of land use, management practices, and innovation in Africa with a view to exploring how multifunctional land uses can alleviate food insecurity and poverty.

Food security and livelihoods in Africa face multiple challenges in the form of feeding a growing population on declining land areas under the impacts of climate change. The overall question is what kind of farming systems can provide resilient livelihoods? This volume presents a selection of existing farming systems that demonstrate how more efficient use of land and natural resources, labour and other inputs can have positive effects on household food security and livelihoods. It examines how aquaculture, integrated water management, peri-urban farming systems, climate-smart agriculture practices and parkland agroforestry contribute multiple benefits. Drawing on case studies from Kenya, Ethiopia, Nigeria and Burkina Faso, contributed by young African scientists, this book provides a unique perspective on multifunctional land use in Africa and illustrates how non-conventional uses can be profitable while promoting social and environmental sustainability. Tapping into the global discussion on land scarcity and linking food security to existing land use change processes, this volume will stimulate readers looking for diversified land uses that are compatible with both household and national food security ambitions.

This book will be of great interest to students and scholars of African development, agriculture, food security, land use and environmental management, as well as sustainable development more generally, in addition to policymakers and practitioners working in these areas.

Elisabeth Simelton is a climate change scientist at World Agroforestry (ICRAF), Vietnam and project leader of the CGIAR research programme Climate Change, Agriculture and Food Security (CCAFS). Her research interests include environmental sustainability issues related to farms, food and the future. Her current work covers landscape adaptation strategies, agroclimate information services and climate policy.

Madelene Ostwald is Senior Researcher in the Gothenburg Centre for Sustainable Development (GMV), Sweden and leader of the Multifunctional landscape theme within the AgriFoSe2030 programme for agriculture for food security. With a focus on land use, forestry and overall multifunctional land use, her research deals with land-based issues related to climate policy, food security, energy, development, and monitoring issues.

Earthscan Food and Agriculture Series

Farming Systems and Food Security in Africa
Priorities for Science and Policy Under Global Change
*Edited by John Dixon, Dennis P. Garrity, Jean-Marc Boffa,
Timothy Olalekan Williams and Tilahun Amede with Christopher Auricht,
Rosemary Lott and George Mburathi*

Consumers, Meat and Animal Products
Policies, Regulations and Marketing
Terence J. Centner

Gender, Agriculture and Agrarian Transformations
Changing Relations in Africa, Latin America and Asia
Edited by Carolyn E. Sachs

A Global Corporate Trust for Agroecological Integrity
New Agriculture in a World of Legitimate Eco-states
John W. Head

Geographical Indication and Global Agri-Food
Development and Democratization
Edited by Alessandro Bonanno, Kae Sekine and Hart N. Feuer

Multifunctional Land Uses in Africa
Sustainable Food Security Solutions
Edited by Elisabeth Simelton and Madelene Ostwald

Food Security Policy, Evaluation and Impact Assessment
Edited by Sheryl L. Hendriks

For further details please visit the series page on the Routledge website:
www.routledge.com/books/series/ECEFA/

Multifunctional Land Uses in Africa

Sustainable Food Security Solutions

Edited by Elisabeth Simelton and
Madelene Ostwald

Routledge
Taylor & Francis Group
LONDON AND NEW YORK

earthscan
from Routledge

First published 2020
by Routledge
2 Park Square, Milton Park, Abingdon, Oxon OX14 4RN

and by Routledge
605 Third Avenue, New York, NY 10017

First issued in paperback 2021

Routledge is an imprint of the Taylor & Francis Group, an informa business

© 2020 selection and editorial matter, Elisabeth Simelton and Madelene Ostwald; individual chapters, the contributors

The right of Elisabeth Simelton and Madelene Ostwald to be identified as the authors of the editorial matter, and of the authors for their individual chapters, has been asserted in accordance with sections 77 and 78 of the Copyright, Designs and Patents Act 1988.

The Open Access version of this book, available at www.taylorandfrancis.com, has be made available under a Creative Commons Attribution-Non Commercial-No Derivatives 4.0 license.

Trademark notice: Product or corporate names may be trademarks or registered trademarks, and are used only for identification and explanation without intent to infringe.

British Library Cataloguing-in-Publication Data
A catalogue record for this book is available from the British Library

Library of Congress Cataloging-in-Publication Data
A catalog record has been requested for this book

ISBN 13: 978-0-367-78542-0 (pbk)
ISBN 13: 978-0-367-24644-0 (hbk)

Typeset in Sabon
by Wearset Ltd, Boldon, Tyne and Wear

Contents

List of illustrations   ix
Author biographies   xii
Preface: background to AgriFoSe2030 and the book project   xiv
Acknowledgements   xv

1 Multifunctional land-use systems – a solution for food security in Africa?   1
ELISABETH SIMELTON, MADELENE OSTWALD AND MOSES OSIRU

2 Nigerian climate-smart agriculture practices with scaling potential   22
SIMON A SHOMKEGH

3 Treating shea trees as crops improves women's livelihoods in Burkina Faso   47
JOSIAS SANOU WITH HUGUES R BAZIÉ AND JULES BAYALA

4 Economic benefits from cassava in peri-urban multiple-cropping systems in Nigeria   61
ANTHONY OJONIMI ONOJA

5 Integrated aquaculture: balancing food production systems and livelihoods in Kenya   78
GERALDINE K MATOLLA

6 What integrated watershed management can deliver for the environment and livelihoods: the Ethiopian experience   96
KASSA TEKA BELAY

7 Smallholder maize-based systems: a piece of the puzzle for
   sustaining food security in Nigeria   115
   JULIUS B ADEWOPO

8 Multifunctional land-use practices in Africa: what else do
   we need to do?   134
   ELISABETH SIMELTON, MADELENE OSTWALD AND
   MOSES OSIRU

   Index   155

Illustrations

Figures

1.1	Multifunctional parkland with crop production supported by characteristic trees	4
1.2	Peri-urban multifunctional land use taking advantage of the height, with green mulch	5
1.3	The conceptual idea behind the Land Equivalent Ratio	13
1.4	Yield of maize, millet, rice and sorghum in Africa and Asia from 1961 to 2017	14
1.5	Geographical location of the six cases. Impacts are listed by category: (1) food security, (2) ecology and (3) socio-economics	17
2.1	Onset and cessation of rainy season in Benue State	24
2.2	Ironwood trees in a maize field in Makurdi, Benue State	28
2.3	A citrus orchard intercropped with cassava in the Benue region	30
2.4	Profitability ratio, gross income to cost, from citrus orchards per farm size ($n=92$)	31
2.5	Farmers' reasons for adopting zero tillage in the Benue region, per cent of respondents ($n=100$, multiple responses were possible)	33
2.6	Profitability ratio, gross income to costs, for zero tillage by farm size ($n=88$)	34
3.1	Shea parkland with millet and maize in Nobéré, Burkina Faso, West Africa	49
3.2	Shea butter processed and traded by a woman in Nobéré, Burkina Faso	51
3.3	Shea nut productivity and harvested area in Burkina Faso from 1980 to 2017	52
3.4	Pruned and unpruned shea trees on a parkland in Nobéré, Burkina Faso, West Africa	54

x *Illustrations*

4.1	Area planted with cassava for Nigeria and Thailand as share of the global area (stacked), and yields (lines) 1985 to 2017	62
4.2	Peri-urban multiple-cropping system with cassava intercropped with yam, and maize in Port Harcourt	68
4.3	Relationship between farm size and cassava yield in multiple-cropping land use by gender (men – M solid symbols, women – W unfilled symbols; $n=150$)	72
5.1	Freshwater aquaculture production between 2005 and 2016 in Kenya	80
5.2	Musuu Farm has an open fish pond with water storage tanks that supply water for drip irrigation for tomatoes and with bamboo trees planted along river bed for preventing soil erosion	85
5.3	Kamuthanga Farm (a) Recirculating aquaculture system with (b) outdoor fish culture tanks close to the agriculture fields with bananas in the background and young tomato plants	87
6.1	Land cover changes in the same area in the Hintalo watershed in 2006 (left) and 2016 (right)	102
6.2	Species evenness, diversity and richness in rehabilitated and non-rehabilitated land	102
6.3	Averaged indicators of soil quality with and without integrated watershed management on dolerite and sandstone watersheds: total nitrogen (per cent), soil organic matter (per cent), mycorrhiza (count), available phosphorous (parts per million)	103
6.4	River water diversion at the Mariam-Shewito watershed, constructed between 2004 and 2006.	105
6.5	Number of months that households ($n=1,676$) were able to cover their own food demand, before and after the integrated watershed management interventions in the Mariam-Shewito watershed	107
7.1	Average maize yields in 2016 in selected countries	116
7.2	Major maize-producing regions in Nigeria	116
7.3	Population, maize area and maize production in Nigeria 1961–2017	118
7.4a	Maize area harvested for Africa's top maize producers 1980 to 2017	120
7.4b	Maize production for Africa's top maize producers 1980 to 2017	120
7.5	A typical mixed maize-based system with okra, soybeans and cowpea on $c.1$ to 2 hectares of farmland in Doguwa, Kano State, in the Sudan Savanna agroecological zone, Nigeria	122

8.1   Animal grazing inside enclosure after harvest. West
      Pokot, Kenya   138

Tables

2.1	Characteristics of interviewed citrus orchard and zero tillage farmers in Benue region	26
2.2	Correlation coefficients for profitability ratio per hectare versus farmer characteristics for orchard and zero tillage	32
4.1	Peri-urban cassava-based multiple-cropping systems	63
4.2	Peri-urban multiple-cropping systems in Port Harcourt ($n=150$)	65
4.3	Farming calendar for the crops in the cassava-based systems in Port Harcourt.	67
4.4	Gross margin of cassava in peri-urban multiple-cropping systems in households headed by men (M) and women (W) in Port Harcourt ($n=150$)	69
6.1	Literature review of selected references	99
6.2	Environmental change and social impact indicators	100
6.3	Distribution of studied watersheds in Tigray, with average elevation	100
7.1	Indicators studied in maize-based multiple-cropping systems in Nigeria	123
7.2	Benefits and potential drawbacks of smallholder maize-based multiple-cropping systems in Nigeria	125
8.1	Example of indicators for assessing the multifunctionality of farming practices	146

Author biographies

Elisabeth Simelton received a PhD in Geography from the University of Gothenburg, Sweden in 2007. She is currently a climate change scientist, based at World Agroforestry (ICRAF) in Hanoi, Vietnam, since 2010. Her research interests focus on land use and land use change related issues, including climate-smart agriculture and on developing farmer-oriented climate services. She is a member of SIANI and Focali Network.

Madelene Ostwald got her PhD in Physical Geography at the University of Gothenburg, Sweden in 2000. She is a senior scientist at the Gothenburg Centre for Sustainable Development (GMV) and theme leader for Agri-FoSe2030 Multifunctional landscapes for increased food security. With a focus on land use, forestry and overall multifunctional land use, her research addresses land-based issues related to climate policy, food security, energy, development and monitoring.

Moses Osiru is the Manager of the Regional Scholarship and Innovation Fund (RSIF) of the Partnership for skills in Applied Sciences, Engineering and Technology (PASET) hosted by the International Centre of Insect Physiology and Ecology (icipe). He is a Plant Pathologist by training. Moses is interested in capacity building in Africa for strengthening the use of science, technology and innovation to achieve Africa's Agenda 2063.

Simon A Shomkegh obtained a PhD in Environmental Management in 2013 and works with the Federal University of Agriculture, Makurdi, Nigeria. He has been in active in teaching, research and outreach service for the past 19 years and currently heads the Department of Social and Environmental Forestry at the University. Simon is an Associate Professor of Environmental Management with current research interests in natural-resources management, ethnobotany and climate resilience building.

Josias Sanou received his PhD in Agroforestry and Ecophysiology from Bangor University, United Kingdom in 2010. He is currently a researcher at Institut de l'Environnement et de Recherches Agricoles (INERA), Burkina Faso. His research interests include the optimization of parkland agroforestry systems, especially in terms of tree-crop interactions and in the context of climate change.

Anthony Ojonimi Onoja is a Senior Lecturer of Agricultural Economics at the Department of Agricultural Economics and Extension, University of Port Harcourt, Nigeria. He obtained a PhD in Resource and Environmental Economics in 2015 and holds one MSc in Agricultural Finance and Project Analysis and one in Resource and Environmental Economics, all from the University of Nigeria, Nsukka. He is the President of the Agricultural Policy Research Network, the Managing Editor of Nigerian Agricultural Policy Research Journal, and the editor of Journal of Agricultural Research. His research interests include resource, development, and environmental economics, agricultural finance, and macroeconomic and agricultural policy analysis using applied econometrics.

Geraldine K. Matolla is a senior lecturer in the Department of Fisheries and Aquatic Sciences and director for Gender Equity and Diversity at the University of Eldoret, Kenya. She holds a PhD in fish parasitology from the University of Eldoret, an MSc in aquaculture from Memorial University of Newfoundland, and a BSc in Wildlife Management from Moi University. She has conducted research under various research projects on pond dynamics, cage fish farming systems and gender issues in fisheries. Her research interests include aquaculture in ponds, cages and tanks, fish parasitology and to integrate community and gender development in aquaculture.

Kassa Teka Belay is an affiliate of the Department of Land Resources Management and Environmental Protection at Mekelle University, Ethiopia. He holds a PhD in Geography from the Department of Earth and Environmental Sciences in KU Leuven in Belgium, an MSc in Tropical Land Resources Management, and a BSc in Land Resource Management and Environmental Management, both from Mekelle University. He is currently working as an Associate Professor and editor for the journal Drylands. His research interests include watershed management, agroforestry, land evaluation, soil fertility and management, and soil survey and mapping.

Julius B Adewopo holds a PhD in Soil and Water from the University of Florida, where he graduated in 2014. He is currently an Associate Scientist at the International Institute for Tropical Agriculture (IITA) where he supports the implementation of a project focussed on advancing smallholder maize-based systems in Sub-Saharan Africa (www.tamasa.cimmyt.org). His research interests encompass delivery of innovations for sustainable natural-resource management, advancing decision support for smallholder farming systems with analytical tools (including geospatial technologies) and engaging stakeholders to co-generate and adopt proven tools and technologies.

Preface
Background to AgriFoSe2030 and the book project

In January 2017 six African young researchers met when participating in an AgriFoSe2030 training course on 'Translating Science into policy and practice' in Nairobi, Kenya. The researchers had different scientific backgrounds, but all shared a research focus on multifunctional land use issues with relation to food security. The in-depth discussions between the researchers were many and apart from all the challenges and issues associated with sustainable land use and food security, it was obvious that the scientists also had research material that demonstrated successes in the field. As an outcome of discussion and debate, they proposed to publish their multifunctional land use case studies as a book. And here we are....

The AgriFoSe2030 programme (Agriculture for Food Security, see www.slu.se/agrifose) is built around a consortium of scientists from Swedish universities. AgriFoSe2030 focuses on sustainable agriculture for increased food security and production. The core activity is translating state-of-the–art science for supporting better policy making and use of improved practices within the agricultural sector, targeting young scientists in the global South as the key agents in this process.

To translate science into policy and practice is hard. The causal link between research-based results and processes outside academia is usually difficult to prove and the timing, language and level of detail is a struggle. Despite these hurdles, the relevance is clear, and the desire exists, particularly within the societies and environments where the authors of this book are working. Therefore, this book can be seen as an important element, and part of the puzzle of using science-based work by scientists in their African context, translated into a format that can be digested by many. We therefore hope that this book can inspire and support the shaping of future policies and practices.

Acknowledgements

We thank AgriFoSe2030 for their financial support as well as the explorative freedom with this project. With open access to this material, it is our wish and ambition to give everyone the opportunity to take part in these cases of multifunctional land use.

The material has been subject to a review process, and we would like to acknowledge and thank the following people for their comments and contributions: Leimona Beira, Delia Catacutan, Peter Bruun Clausen, Karl-Erik Eriksson, Sabine Henders, Magnus Jirström, Cajsa Lithell, Anders Malmer, Eskil Mattsson, Rachmat Mulia, Kelvin Shikuku, Nhuong Tran and Ingrid Öborn. We are also thankful for the language support from Paulina Essunger and editorial assistance from Cuong Manh Nguyen, Loan Thanh Pham and Ha My Tran.

Elisabeth Simelton, Hanoi, Vietnam
and
Madelene Ostwald, Göteborg, Sweden
March 2019

1 Multifunctional land-use systems – a solution for food security in Africa?

Elisabeth Simelton, Madelene Ostwald and Moses Osiru

What is multifunctional land use?

Multifunctional land use is based on systems that are managed with the goal of producing more than one product or service. The products can be, for instance, grains, fodder, timber, firewood, biofuel, fruits or flowers, while the services can be water infiltration, wind breaks, microclimate regulation, carbon storage, erosion control, groundwater recharge or soil conservation, among others. Mander *et al.* (2007) describe landscapes as multifunctional through their simultaneous support of habitat, productivity, regulatory, social, and economic functions. Heterogeneity (diversity), they noted, is a basic attribute of landscapes, and this heterogeneity implies the capacity of the landscape to support various and sometimes contradictory functions simultaneously.

The term 'multifunctionality' was coined by the Organisation for Economic Co-operation and Development (OECD) and the European Union (EU) in the early 2000s and grew from a debate that aimed at reforming the European Common Agricultural Policy from conventional production towards a rural development orientation (Wiggering *et al.* 2006). Conventional agriculture in the western countries typically refers to monoculture that uses synthetic chemicals and other agricultural inputs, where the primary objective is market-oriented (USDA 2015). The term 'multifunctional' gained further credence as the World Trade Organization reduced trade barriers and production-based farming subsidies (COM 2002 in Wiggering *et al.* 2006). These actions were a reaction to the fact that public environmental goods were undervalued and therefore misused (Wiggering *et al.* 2016). Hence, the transition from conventional to multifunctional agriculture centred around two parallel types of incentives that aimed at: (i) having farmers or land users reduce negative environmental effects, (ii) having consumers or authorities create markets and demand for diverse rural products and services, sometimes with the help of subsidies, penalties, or payments (Vereijken 2003). Consequently, multifunctional land use brought together planning-concept perspectives (Vreeker 2004) and problem-solving perspectives (Wiggering *et al.* 2006).

In a European perspective, conservation of nature, agricultural landscapes and cultural heritage values are associated with human and animal health and well-being, tourism and recreation, which can contribute to agricultural or rural employment (OECD 2001). In Europe, the inclusion of rural employment and food security in the discussion of multifunctionality has been controversial. Rural employment in agriculture is typically viewed as an input rather than a non-commodity output of agriculture or an externality. However, rural employment can also have societal impacts that can be considered externalities, such as slowing migration from rural to urban areas (OECD 2001).

In the context of developing countries in the South, the interactions between food security, rural livelihoods and societal outcomes are noticeable. In the light of population growth and climate change impacts, food security is becoming more than a basic component of health and well-being for achieving or maintaining any of the other functions. Transitions between conventional production-oriented land uses and multifunctional ones involve the loss or integration of more rural functions at any scale, including (adapted from Vereijken 2003):

- **production:** food, feed, fibre, fuelwood, biofuel, timber, flowers;
- **environment:** windbreaks, erosion control, groundwater recharge;
- **nature and landscape:** biodiversity, habitat, agricultural and cultural heritage;
- **climate:** carbon storage, microclimate regulation;
- **work and income:** rural employment, urban migration; and
- **health and well-being:** food security and nutrition, agro-tourism, recreation.

Drivers of multifunctional land use

Although the origin of the term 'multifunctional land use' is related to European-centred conservation, people around the world live in multifunctional landscapes and practice multifunctional land uses as part of their livelihoods. The drivers of various types of multifunctional land uses can be divided into, but are not limited to and may be combinations of, for example:

- traditional systems for subsistence;
- scarcity of food, land, or labour;
- innovation for improved production;
- policies for specific goals, such as conservation goals or climate mitigation; or
- market demands, such as ecological farming, niche farming.

Traditions

Many traditional land uses have developed over long time periods as interactions between environmental functions and cultural benefits. Shifting cultivation is one such example, which has existed in nearly all agroecological zones at some point in history, primarily for subsistence farming. The system typically includes a rotational slash-and-burn practice, with fallow periods to regain soil fertility and a sequence of crops that responds to declining soil fertility during the cultivation phase. Eventually, with land scarcity, the fallow periods become shorter and soil fertility declines, and the shifting cultivation systems can no longer sustain production. These traditional farming practices are effectively the results of accumulated indigenous knowledge, culture and adaptations passed on from generation to generation, before scientific agricultural research and extension systems gained ground.

The term 'agroforestry' was coined in the 1970s, as a collective name for practices in which farmers were deliberately planting or keeping trees on agriculture land (Nair 1993). However, the general practice was thousands of years older, as farmers learned early on that there was gain from multiple benefits, products and services by mimicking natural-forest systems with multiple canopy layers, keeping animals close to trees, or growing homegardens. For example, hedges and trees can serve as demarcation, as is seen with enclosures (Figure 8.1 in Simelton, Ostwald and Osiru Chapter 8) or exclosures used to separate livestock from cultivated land (Woodhouse 2003), or as habitat for pollinators in vineyards or similar production systems. Agroforestry can also be applied to enhance biomass production, stabilize soil or conserve water in natural vegetation or human-made productions systems, such as parklands. Parklands as traditional multifunctional land-use systems exist throughout the Sudano-Sahelian part of Africa (Karlson 2015) and are the setting for our chapter (Sanou Chapter 3) on shea production (the nut from the tree *Vitellaria paradoxa*) in Burkina Faso (Figure 1.1). In these systems, the regular production of one or more agricultural crops is supported by scattered trees that supply additional products such as fodder, fruits or fuel wood while enhancing crop productivity through improved water retention, soil structure and fertility.

Scarcity

Homegardens and backyards can serve as a food shelf containing diverse short-term vegetables and fruits that supply daily diets with important micro-nutrients, especially where scarcity of land or income is an issue. In urban environments, landless people use unused patches or wasteland, sometimes with unclear land entitlements, to feed themselves (Figure 1.2). One such case is from Nigeria (Onoja Chapter 4).

4 *Elisabeth Simelton* et al.

Figure 1.1 Multifunctional parkland with crop production supported by characteristic trees.
Photo credit: Ostwald 2017.

As livestock are often kept near homes, manure can be recycled for compost to restore soil fertility. Moreover, fish ponds near homesteads are also a way to store water, recycle household waste and reduce food scarcity in a multifunctional setting, which is described in our chapter on fish farming in Kenya (Matolla Chapter 5). Rice-fish cultivation has been practised for millennia, predominantly in Asia and some parts of Africa. Fingerlings are introduced into paddy rice fields, or fish enter naturally when rivers flood the fields. The fish feed on molluscs, insects, or waste products, and will do the weeding and natural fertilization without affecting rice yields. Besides being land-use efficient, this practice reduces farmer labour inputs for maintenance (Halwart and Gupta 2004).

Figure 1.2 Peri-urban multifunctional land use taking advantage of the height, with green mulch.
Photo credit: Simelton 2018.

Innovations

New methods and ideas to increase food production can bring about multifunctional landscapes. Compared to drivers of traditional practices and scarcity, innovation incorporates components of exploration or testing. Adding a new practice, crop or management to existing structures can enhance production and thus benefits or revenues. Homegardens are among the least regulated land-use systems; policies have had limited influence on designs and content. Therefore, these gardens become sites for land users' experimentation and domestication of species and are also great biodiversity banks (Mulia *et al.* 2018).

Innovations, in this context, refer to technical solutions or products as well as processes, such as collective action and social learning, that foster transitions towards sustainable agriculture and multifunctional landscapes (Pigford *et al.* 2018). For example, the climate-smart village concept serves to establish communities with climate-smart agriculture practices for upscaling (Aggarwal *et al.* 2018). The documentation found in the chapter on climate-smart agriculture (Shomkegh Chapter 2) in Nigeria exemplifies the importance of other social processes than those based on climate-smart villages.

In a global context, despite being seen as a geographical area of great potential the African continent has not been able to adequately make use of farming innovations as well as have other developing regions (Meijer *et al.* 2014). In contrast, some argue that Sub-Saharan African rural landscapes have been influenced by external international agendas, as portrayed by the Green Revolution's promotion of monocultures, and that this resulted in the loss of smallholders' multifunctional livelihoods (Dawson *et al.* 2016).

Innovative practices can spread between practitioners (Weltzien and Christinck 2017) or be picked up and extrapolated by other agents, such as agricultural advisory service providers (extension) or development and research organizations, which we see in our chapter of integrated maize production in Nigeria (Adewopo Chapter 7). Often private capital and investment can boost the uptake and co-creation process. An example of innovation is the work of VI-agroforestry (a Swedish development organization focusing on planting trees and improving livelihoods) in eastern Africa. The chapter on fish farming also demonstrates how innovations are dependent on risk-takers to lead the process. We foresee that some urban areas will lead future technological innovations in multifunctional farming, such as three-dimensional or vertical farming in new settings.

Policy

Policy drivers towards multifunctional land uses are often based on international or national commitments, involving subsidies in one way or another. National strategies involving multifunctional land uses are now beginning to take shape, such as agroforestry strategies in India and ASEAN member states (Catacutan *et al.* 2018). When the European Union agreed to refer to different types of evenly and unevenly distributed woody vegetation as agroforestry, the products and services that this land use contributed to rural development and environmental resilience could be better estimated. With a joint definition and evident contributions to global commitments on biodiversity and climate mitigation, agroforestry was suddenly visible in policy and eligible for support measures, such as agroenvironmental payments (Mosquera-Losada *et al.* 2016).

One early policy-driven process was seen in Vietnam in the 1970s and 1980s, where traditional multifunctional land uses were reintroduced after the war. Land allocation programmes for homegardens, fish ponds, some livestock and a mixed forest were introduced. The policy aimed to ensure household food security and contribute to reforestation targets and a shift from previously nomadic and semi-nomadic livelihoods in increasingly degraded forests. Reforestation activities were funded with bilateral aid and loans (Catacutan *et al.* 2016). Another policy with multifunctional land use is the Brazilian Low-Carbon Agriculture Plan starting in 2010. The climate-driven plan is a credit initiative that provides low-interest loans to farmers who want to implement sustainable agriculture practices. Despite its criticized set-up and impact (Newton *et al.* 2016), the land-use changes that are emerging are integrated crop-livestock-forestry systems, no-till farming, restoration of degraded forests and pastures, as well as manure management, all with the purpose to reduce greenhouse gas emissions and supply agricultural products and ecosystem services.

In many developing countries, the funding of the 'green' rural sector has shifted to global financial mechanisms. The Global Environment Facility was established in 1992 to address environmental problems and is a financing mechanism for the Conventions on Biodiversity (CBD), Desertification (UNCCD) and Climate Change (UNFCCC). In addition, in 2009 the Green Climate Fund was established and focuses on climate adaptation and mitigation activities within the UNFCCC framework. Other mechanisms within the UNFCCC, such as the Clean Development Mechanism, Reducing Emissions from Deforestation and Forest Degradation (REDD+), and strategies in countries' Nationally Determined Contributions (NDCs), also show the link between policy drivers and multifunctional land uses. Even if the former mechanisms have had less representation in Africa, NDCs exist for all African countries and have a strong focus on land use and forestry. Further, the least developed African countries are particularly keen to account for agroforestry in their NDCs. One example of policy-driven land use is found in our chapter from Ethiopia (Teka Chapter 6), where watersheds were targeted for rehabilitation and ecosystem improvement through a number of interventions.

Market

Increasingly, markets determine the value of land and what is grown on the land. Where urbanization increases, staple crops become too expensive and eventually disappear, while some land patches are used to meet the demands of middle-class markets or high-end restaurants. This creates opportunities for new types of scattered multifunctional land uses. For example, urban and peri-urban agroforestry are emerging as new multifunctional practices that integrate rural and urban development (Borelli *et al.* 2017). Niche farming offers a targeted product and/or services for well-defined market segments, such as online sales or agro-tourism. Typically, it focuses on one core activity with few fresh or processed products, such as organic vegetables or honey. Such businesses require not only land but also entrepreneurial skills and may involve the transformation of conventional farms or initiate as small start-ups and contribute to multiple rural values (Anzaku and Salau 2017; Pigford et al. 2018). The chapter on fish farming in Kenya illustrates some of the challenges in starting up niche farming.

Marketing, branding and certification schemes involving multifunctional land uses are also a growing segment. Sensitive to higher temperatures, arabica coffee plants are normally grown at higher altitudes (Rahn et al. 2018). As temperatures continue to increase, traditional ways of growing coffee plants under tree canopies are therefore regaining popularity. The shade tree regulates the microclimate, which also improves the quality and marketing of coffee (Hernandez-Aguilera et al. 2018).

We remind ourselves of the need to view the interactions of multiple functions beyond their roles in the field, to the landscape scale. A common argument is the need to intensify production somewhere in order to save land or avoid environmental degradation elsewhere. A modelling example from the Democratic Republic of Congo shows that this theory may not hold, the renting out of agricultural land was driving deforestation (Phelps et al. 2013). Two chapters from Nigeria (Onoja Chapter 4; Adewopo Chapter 7) suggest somewhat similar trajectories.

Global extent

The lack of a common definition of multifunctional land uses makes it hard to assess, quantify or estimate the importance of the practice. One reason is that the term encompasses diverse practices and systems, such as agroforestry, homegardens, parklands, different types of integrated cropping systems, trees outside of forests, and urban and peri-urban farming. Scholars (for instance, Wilson 2008) have also argued that there is a lack of research around multifunctional land uses and that one way forward would be to acknowledge the spatially complex nested hierarchy that the practice contains, so that the only starting point is 'on the ground' of that particular practice and where the decisions are being made. The quantification problem is also seen in agricultural statistics, which report on single crops rather than on the combinations in which they are grown. Ultimately, without definitions, there are no budget lines for public spending.

One option with the potential to bypass this challenge and allow for quantification is agroecological zoning (Leff et al. 2004). Leff and colleagues (2004) developed an Agricultural Commodity Diversification Index (ACDI) per pixel, in order to demonstrate the importance of other food crops beyond the 'big three' of wheat, maize and rice. This index could be the basis for a more integrated assessment of diverse agricultural systems.

A more indirect impact on the global extent of multifunctional land use is an approach by Zomer et al. (2016), who used remote-sensing data to assess agricultural land with trees. The global carbon stock contribution of these multifunctional land-use types was studied for the period between 2000 and 2010. MODIS satellite data revealed that out of the world's 2,200 million hectares of agricultural land in 2010, 43 per cent had at least 10 per cent tree cover. The amount of tree-covered agriculture land in Africa is 260 million hectares, land that in general showed a declining carbon stock over the ten-year period. Apart from their main conclusion that these lands hold great carbon sequestration potential, there are positive side-effects of improved soil water-holding capacity and increased crop productivity.

Another option to better quantify the global extent of multifunctional land uses is through Earth-observing satellite data and geospatial technologies and tools, which are becoming increasingly available and accessible. Open source tools, such as Collect Earth (http://collect.earth/), developed by the Food and Agriculture Organization of the United Nations, and the SERVIR programme (www.servirglobal.net) for monitoring land use and land-use changes, will also be of help in the documentation of multifunctional land uses.

Trade-offs, drawbacks and benefits

Farmers' trade-off calculations between specializing in one crop or integrating several can often be related to the value chain and benefits of scale, even when farms are small. Monoculture is often perceived as easier to manage in terms of the utlization of inputs, planting, maintenance up to harvest, post-harvest processing, and sale of products. First, this means that agriculture equipment and agrochemicals can be applied without risk of damaging other trees or crops on the field. Second, seasonal labour can be hired to cover peaks. In contrast, multifunctional land-use practices may be hampered by the absence of commercial actors for the diversified production, contract farming or uncertain tenure situations. This is described in the chapter from Burkina Faso (Sanou Chapter 3), which describes shea production from *Vitellaria paradoxa* trees in the parkland system as underutilized.

Diversified farming systems typically depend on daily labour inputs, requiring somebody to stay on the farm. This should be seen in contrast to off-farm jobs that may render additional cash incomes. However, integrating higher-value crops may provide livelihood options for those who choose to, or must, stay on the farms. Further, the selection of crops must consider the possibilities that roots and growth may cause competition for water, nutrients or shade. Three chapters about climate-smart agricultural practices (Shomkegh Chapter 2) and cassava-based (Onoja Chapter 4) and maize-based systems (Adewopo Chapter 7) describe how farmers try to overcome these challenges. In addition, if new knowledge is required, such as planting or landscaping techniques, a functional extension system, input support and farmers' own or public investments may be costly and become a bottleneck. The example of integrated watershed management from Ethiopia raises these points (Teka Chapter 6). Therefore, unless farmers learn from each other, participatory community processes to identify new multifunctional systems that build on existing experiences have a greater chance of adoption (Aggarwal *et al.* 2018; Duong *et al.* 2016).

Contrasting monoculture and multifunctional land uses may be counterproductive for several reasons. First, such comparisons tend to fall into traps of conventional economic reasoning, where externalities and non-monetary values are unaccounted for. Second, the bias towards monocultures in policies, extension, statistics, and experimental research makes it difficult to counter-argue with relevant evidence (see Mattsson *et al.* 2018). Conversely, multifunctional land use is hampered by its broad and undefined scope that can incorporate all or nothing and is sensitive to context. Farming systems that are diverse, flexible, and context-specific are thus viewed as 'difficult' to implement and assess in policy targets and outcomes. The multifunctional characteristics typically also involve several institutional bodies – energy, forest, agriculture, water, environment departments – who each have their own priorities. This institutional and ownership status can be a drawback in developing multifunctional land-use systems, which is seen in this book. In Burkina Faso, trees belong to the land owner while the crops belong to land users, which caused conflicts rather than co-benefits (Sanou Chapter 3). In Nigeria, agricultural intensification caused forestry degradation (Adewopo Chapter 7). In Kenya, gaps in the extension service failed to recognize fish farming as a prosperous option for small-scale farmers (Matolla Chapter 5). This difficulty in assessing productivity of multifunctional landscapes has often led to the assumption that small farms are not as productive as large farms. However, we know from Asia that farm size is not the key determinant of productivity.

When farmers mix two or more species, they do this because they see benefits of multifunctional systems that outweigh those of monoculture. Farmers have traditionally been viewed as risk averse, therefore diversification of crops has always meant diversification of risks. With farming enterprises becoming risky due to more variable climatic patterns, adding trees in the landscape can reduce negative weather impacts (adaptation benefits) and result in shorter economic recovery periods after natural disasters (Simelton *et al.* 2015). This means making use of environmental functions such as microclimate regulation, improving light-nutrient-water efficiency and improving soil status. Peanut (*Arachis hypogaea* L.) is an example of a cover crop that can be intercropped with cassava or maize, and as a legume it also makes nitrogen available to plants, thus reducing the need for added fertilizers. Many of these practices contribute to sequestering carbon or reducing greenhouse gas emissions from land uses. When the global potential of the carbon pool of multifunctional land uses is estimated (Zomer *et al.* 2016), the motivation for countries to account for agroforestry in

Nationally Determined Contributions may increase. For example, a majority of the 56 countries that had accounted for agroforestry in their 2015 contributions recognized both adaptation and mitigation co-benefits (Rosenstock *et al.* 2018). Agroforestry can be considered a reforestation stage and a practice that avoids deforestation or forest degradation (leakage). Specifically, when assessing homegardens in Sri Lanka, Mattsson *et al.* (2014) found that smaller gardens had more biomass (and hence more carbon) per unit area than larger gardens. Evidence from Vietnam suggests that in areas with severe natural-forest degradation, homegardens may be an important source for local biodiversity conservation (Mulia *et al.* 2018), besides a diverse source of nutrients. A rigorous global review of homegardens globally shows both that multifunctionality benefits are well represented and that there is a need to further understand economic and non-economic values of homegardens related to women's livelihoods, nutrition, and education as well as to post-conflict solutions (Galhena and Maredia 2013).

Assessing multifunctional land use and food production

There are two problems with how we are taught to measure farm productivity. First, conventional farm productivity is evaluated based on summed-up monoculture yields, rather than assessing the nutritional value, profits and multiple ecosystem functions of all species in combination. Second, the conventional agricultural view is based on two-dimensional production systems, where the ambition is to maximize the output per unit area such as yield per hectare, while multifunctional systems allows planning for production in both the horizontal and the vertical plane, such as multi-storey plantations (Figure 1.2). The shift of units is not impossible to overcome, but it is still a shift in mind-set to one that is closer to forestry than agriculture.

Agricultural research and climate impact-food security studies are often preoccupied with closing yield gaps and variability. Smallholder farmers' yields rarely reach the levels they would under perfect conditions of timing, water and nutrients. As improved crop varieties have a narrower window of optimal conditions, exploring how to close such yield gaps could make attainable contributions to global food security levels (Evans and Fischer 1999; Lobell *et al.* 2009). Yield-gap studies are useful in that they help us identify inefficiencies in management. However, both simulated potential yields and experimental yields can be deceptive as the type and number of limiting factors at the farm level are more diverse. Hence, a more feasible priority is lifting the average farmer closer to the maximum farmer's yield (Lobell *et al.* 2009). When yields are becoming more variable, yield losses could be avoided by shifting to more stress-tolerant crops, for instance by shifting from maize to sorghum (Lobell *et al.* 2009), or millet, which are sometimes more nutritious.

What tends to be forgotten in these kinds of climate-crop model studies is that yield gains could also be achieved through the positive interactions between trees and crops that make use of environmental functions.

- **Reduce the variability of yields** by providing buffers against weather-related stress. Canopies, stems or roots of one crop protect another crop against wind, sunshine and soil erosion during periodic or constant risk of stress. Different root lengths avoid crops competing for soil moisture at the same depth, and their root systems improve the stability of both plants and soil. Shade reduces the temperature below the canopies, which lowers the evaporative demand directly from the soil surface and helps plants make better use of soil water via evapotranspiration. Temperature and soil moisture also regulate the stomata and photosynthesis functions. This translates directly into crop growth as stressed plants are more prone to disease and pest.
- **Increase yields** by modifying nutrient-limiting conditions. Adding legumes, or so-called fertilizer trees with nitrogen-fixing roots, helps crops take up nutrients.
- **Improve economic resilience** – diversified systems reduce the risk of losing the whole harvest to natural or economic disasters. The advantage of spreading risk across the year needs to be considered in relation to trade-offs on labour inputs, if the farm depends on seasonal job migration or hired labour.
- **Store more carbon in trees and soils** – while contributing to climate change mitigation, the economic benefits, such as opportunities to generate additional income or benefits to households through carbon credits or schemes with payments for ecosystem services, are likely to be more motivating for smallholder farmers.

A critical measure of multifunctional land use needs to capture tree-crop interactions to demonstrate land-use efficiency of diversified production and yields. The Land Equivalent Ratio (LER) compares the relative areas required to produce a given yield from two crops in a) monoculture systems versus b) an intercropping system (Figure 1.3). The ratio is calculated as the intercrop production divided by the monoculture production, for each product and per hectare. For example, a LER of 1.4 means that production equivalent to that on one intercropped hectare would require 1.4 hectares if the components (trees and crops) were grown separately, or that intercropping produces 40 per cent more than monocropping. Depending on the purpose, this measure can be used for comparing all products, only the commercial products, or the total biomass produced on one plot. The ratio helps to optimize spacing and thinning schemes for timber trees (Borrell *et al.* 2005). In assessing the competition between plants in greenhouses, Taha and El-Mahdy (2014) demonstrated that the

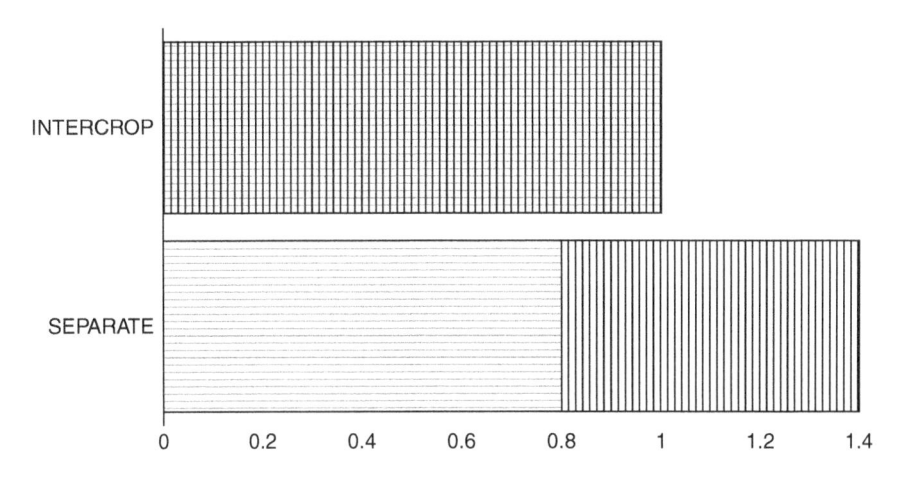

Figure 1.3 The conceptual idea behind the Land Equivalent Ratio.
Source: Modified from Mead and Willey (1980).

LER could capture both which combination of crops achieves the highest yield advantage and the actual magnitude.

To assess sustainable multifunctional land uses, Wiggering *et al.* (2006) propose weighting the economic and ecological utilities. They developed production possibility curves by defining indicators of social utility that merge both commodity outputs, which are paid for on the market, and non-commodity outputs, which are public goods, typically environmental functions such as soil and the climate properties of a landscape. The highest achievable value of social utility on the curve is called a welfare optimum, which represents the maximum production of commodity and non-commodity outputs.

Rethinking farming systems

Within one generation, Africa's population is expected to double, reaching 2.5 billion by 2050. Over the same period, the share of urban citizens will increase from four out of ten, to six. Adding to this, climate change impacts will increase heat and water stress. Here, we outline five concrete production factors that future generations of scientists, policy makers and planners will need to consider when handling the massive challenge.

1 **Land.** Africa's total current cropland is 270 million hectares (FAOSTAT 2019), or 9 per cent of the continent. By 2050, each hectare of cropland will need to support twice as many people, corresponding to an increase from 70 to 140 persons per hectare on average. This may be done by (i) producing more per hectare, for

example by improving the LER; (ii) monitoring that solutions that cause the conversion of other land uses do not trigger unwanted processes, such as deforestation or grassland conversions with wildlife and habitat destruction; and (iii) managing land tenure to avoid further land fragmentation.

2   **Technology.** African key staple crop productivity does not reach global average rates. For example, current yields of maize, millet, and rice are only half of those in Asia (Figure 1.4). This yield gap motivates consideration of how to significantly increase yields in Africa, which is more likely than in other continents. Methods range from indigenous methods to genetic modification and high-tech infrastructure. For example, intercropping indigenous fertilizer trees such as *Faidherbia albida* in certain parklands systems can increase crop yields, such as barley (Hadgu *et al.* 2009). For smallholders, versatile tools and equipment for diverse crops are important in order not to lock poor farmers into monoculture systems. The feasibility of the required productivity increases depends on multipurpose water-harvest and water-saving technologies that support human and agricultural needs without depleting groundwater resources.

3   **Labour.** Of the growing population, the majority will live in cities and not be involved in on-farm food production. In most countries across the world, urban migration results in age, gender and income biases, where the oldest and youngest generations and more women than men are staying in rural areas, possibly depending on remittances from

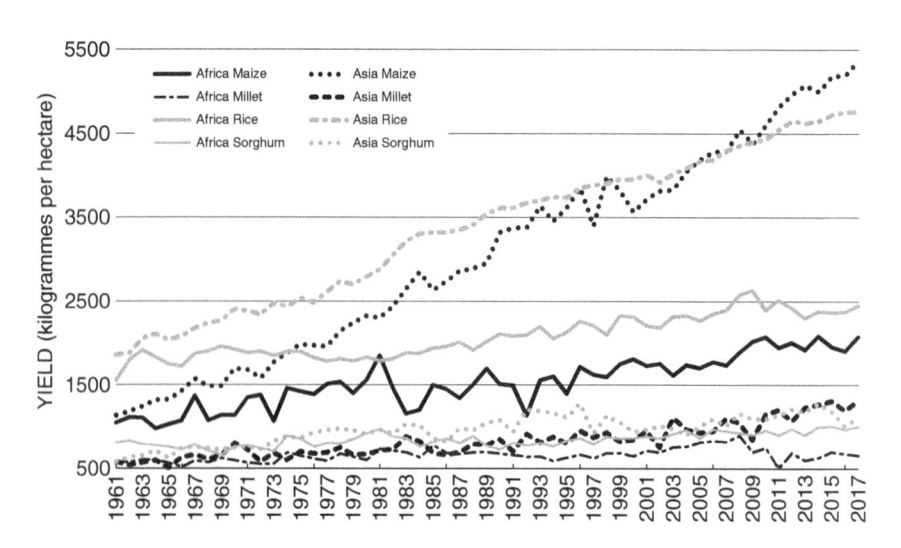

Figure 1.4 Yield of maize, millet, rice and sorghum in Africa and Asia from 1961 to 2017.

Source: FAOSTAT 2019.

their urban relatives (FAO 2016; McKay 2005; Mohapatra and Ratha 2011). This can result in different rural labour scenarios: some remain farmers, such as in the peri-urban agriculture case in Nigeria (Onoja Chapter 4); some do off-farm agriculture work, illustrated in Kenya by the fish-farming chapter (Matolla Chapter 5); and some leave agriculture for non-farm activities. It is relevant to ask what type of farmer will choose which scenario and what demographic and land-use consequences this may cause. If women make up a large part of the rural labour force, do any traditions restrict women from certain equipment or crops? The fish-farming chapter exemplifies how changes in fishery technology pushed women out of traditional income sources. In contrast, the parkland chapter from Burkina Faso exemplifies how gendered traditions can be turned into opportunities (Sanou Chapter 3).

4 **Economy.** Income inequalities in Sub-Saharan Africa are among the highest in the world. The world's three highest Gini coefficient values, all above 0.60, indicating high inequality, are found in South Africa, Namibia and Botswana. The four countries described in this book range between 0.33 in Ethiopia and 0.48 in Kenya (WB 2019). Managing the trends in income disparities will be required to ensure food security, especially for those who no longer grow their own food. For example, Engel's Law relates food insecurity to the share of household income spent on food, thus poor households are more sensitive to food price inflation (Tschirley *et al.* 2015). Smith and Subandaro (2007) considered households that spend more than half their income on food medium food insecure, and those spending more than three-quarters very vulnerable, meaning food insecure. Solving this dilemma is delicate, as the push for cheaper food that low-income consumers can afford risks making farmer income lag. Food-secure farmers are more likely to take in new extension information and adopt new practices (Ragasa and Mazunda 2018); this may be why food-for-work programmes, such as those in Ethiopia, attract a certain type of farmer and not others, which is described to some extent in the chapter on integrated watershed management in Ethiopia (Teka Chapter 6).

5 **Policies and governance.** In their review of the twentieth century African smallholder policies, Birner and Resnick (2010) show how the diversification of the actors involved has influenced policy formulation more than policy implementation. In particular, many countries have undergone shifts towards democracy and multi-party systems and decentralization. Farmers are becoming increasingly organized and connected to the internet, and the private sector – including supermarkets and multinational companies – have gained influence over what is grown. Finally, the answer to the question of whether smallholders benefitted from the structural adjustment programmes is complex. The answer depends on whether countries decided to spend subsidies on inputs for farmers or on food prices for consumers. Birner

and Resnick further distinguished between food crops and crops for export and suggest that richer farmers may have benefitted from trade liberalization policies on food crops. The global food price crisis in 2008 put those policies to a real-time test, when people in many countries no longer could afford to buy food, triggering riots. In response, some countries did nothing, some subsidized consumers, others subsidized farmers and some banned exports or ran into debts. This thread is further discussed in the concluding chapter (Simelton, Ostwald and Osiru Chapter 8). The degrees to which governments interfere in agriculture, markets and trade situations also vary, as shown by the historical contexts described in the chapters from Nigeria and Kenya.

Paradigm shifts take place when both the policy and development partner agenda converge on more integrated policies, such as the Sustainable Development Goals and climate change outcomes. Opportunities for multifunctional land uses appear more appealing in the context of rural transformation, which focuses more on rural-urban linkages and where agriculture has direct and indirect roles to play.

The African case studies in this book

This book draws experiences from six case studies on multifunctional land use across Africa, including climate-smart agriculture (Nigeria, Shomkegh Chapter 2), women's livelihood and shea trees systems (Burkina Faso, Sanou Chapter 3), peri-urban cropping systems (Nigeria, Onoja Chapter 4), fish farming (Kenya, Matolla Chapter 5), integrated water management (Ethiopia, Teka Chapter 6), and maize-based cropping systems (Nigeria, Adewopo Chapter 7) (Figure 1.5). It is recognized that the book presents six land-use cases from a continent of 3,000 million hectares. However, the book does demonstrate that there are success stories out there that, in the right context, including policy support, could significantly impact the continent. Importantly, a common trait from the stories was that the main driver towards multifunctional land-use practices was an increased demand for food. The demand for food was associated with population increase, low yields, a large share of smallholder farmers with fragmented lands, low incomes and investment capacity, uncertain tenures and vulnerability to climate change.

Each of the six case studies shows an innovative improvement to difficult challenges that Africa is facing. The examples cover a range from low-cost adaptation of traditional systems, to investment demanding modernized solutions. The land uses, multifunctional, per definition, have all resulted in more than one product and service that have contributed to improved food security and livelihoods. We hope that the cases will inspire more debate, enhanced documentation, new testing grounds and hence better development of new multifunctional land uses.

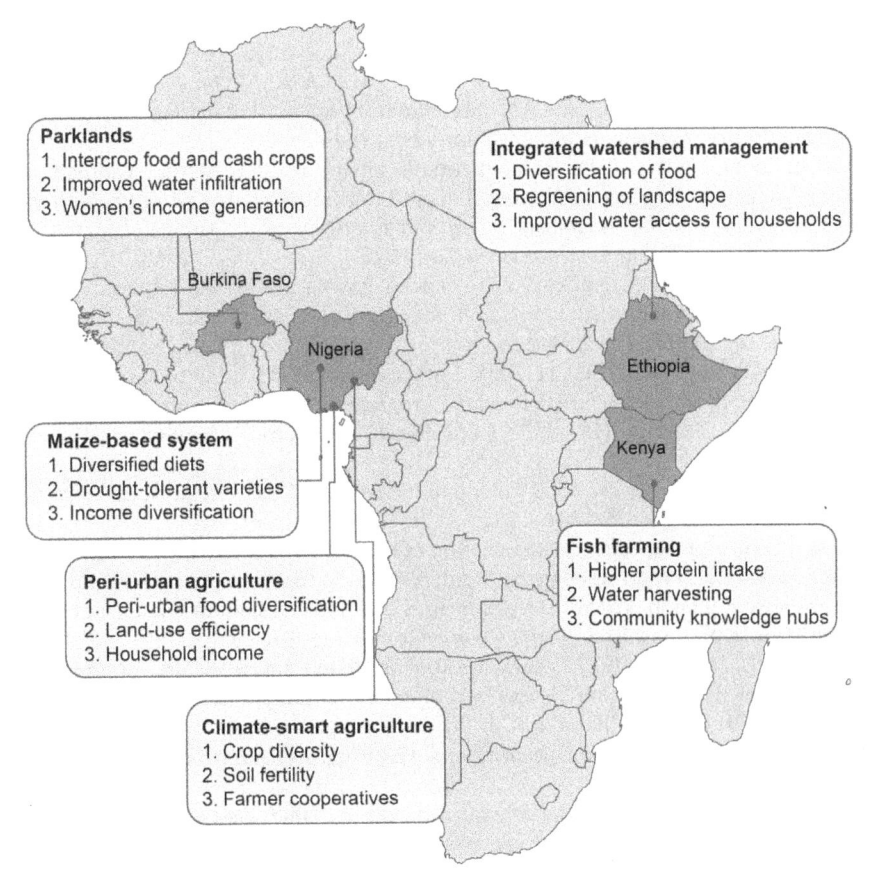

Figure 1.5 Geographical location of the six cases. Impacts are listed by category: (1) food security, (2) ecology and (3) socio-economics.

References

Aggarwal P, Jarvis A, Campbell B, Zougmoré R, Khatri-Chhetri A, Vermeulen S, Loboguerrero A, Sebastian L, Kinyangi J, Bonilla-Findji O *et al*. 2018. The climate-smart village approach: framework of an integrative strategy for scaling up adaptation options in agriculture. *Ecology and Society* 23(1):14.

Anzaku TAK, Salau ES. 2017. Niche marketing potentials for farm entrepreneurs in Nigeria. *Journal of Agricultural Extension* 21:136–142.

Birner R, Resnick D. 2010. The political economy of policies for smallholder agriculture. *World Development* 38(10):1442–1452.

Borelli S, Conigliaro M, Quaglia S, Salbitano F. 2017. Urban and Peri-urban agroforestry as multifunctional land use. *In*: Dagar J, Tewari V (eds). *Agroforestry*. Springer. Singapore. P 705–724.

Borrell T, Dupraz C, Liagre F. 2005. Economics of silvoarable systems using LER approach. AGROOF, Société coopérative et participative spécialisée en agroforesterie. Montpellier, Paris. https://agroof.net/agroof_dev/documents/safe/Economics_silvoarable_systems_LER_approach.pdf.

Catacutan D, Finlayson R, Gassner A, Perdana A, Lusiana B, Leimona B, Simelton E, Oborn I, Galudra G, Roshetko J et al. 2018. *ASEAN Guidelines for Agroforestry Development*. ASEAN Secretariat. Jakarta, Indonesia. www.worldagroforestry.org/region/sea/publications/detail?pubID=4392.

Catacutan D, Nguyen T, Do TH, Simelton E, Hoang T. 2016. Shifting Cultivation in Vietnam – Impacts of Various Policy Reforms *In*: Cairns M (ed.). *Shifting Cultivation Policies. Balancing Social and Environmental Sustainability*. CABI. Wallingford, UK and Boston, USA. Chapter 25.

Dawson N, Martin A, Sikor T. 2016. Green Revolution in Sub-Saharan Africa: implications of imposed innovation for the wellbeing of rural smallholders. *World Development* 78:204–218.

Duong MT, Simelton E, Le VH. 2016. Participatory identification of climate-smart agriculture priorities. CGIAR Research Program on Climate Change, Agriculture and Food Security (CCAFS). Copenhagen, Denmark. http://hdl.handle.net/10568/75542.

Evans L, Fischer R. 1999. Yield potential: its definition, measurement, and significance. *Crop Science* 39:1544–1551.

[FAO] Food and Agriculture Organization of the United Nations. 2016. Migration, agriculture and rural development. Addressing the root causes of migration and harnessing its potential for development. Food and Agriculture Organization of the United Nations. Rome. 20p. www.fao.org/3/a-i6064e.pdf.

FAOSTAT. 2019. FAOSTAT database (online). Food and Agriculture Organization of the United Nations. Rome. www.fao.org/faostat/en/#data.

Galhena DH, Freed R, Maredia KM. 2013. Home gardens: a promising approach to enhance household food security and wellbeing. *Agriculture & Food Security* 2(8):48–62.

Hadgu K, Kooistra L, Rossing W, van Bruggen A. 2009. Assessing the effect of *Faidherbia albida* based land use systems on barley yield at field and regional scale in the highlands of Tigray, Northern Ethiopia. *Food Security* 1(3):337–350.

Halwart M, Gupta MV. 2004. Culture of fish in rice fields. Food and Agriculture Organization of the United Nations and the WorldFish Center. Rome, Italy and Penang, Malaysia. 83p. www.slideshare.net/fullscreen/pd81xz/ot-b-269/1.

Hernandez-Aguilera J, Gómez M, Rodewald AX, Rueda X, Anunu C, Bennett R, van Es H. 2018. Quality as a driver of sustainable agricultural value chains: the case of the relationship coffee model. *Business Strategy and the Environment*. Special issue: challenges of the sustainability transition in global trade 27(2):179–198.

Karlson M. 2015. Remote sensing of woodland structure and composition in the Sudano-Sahelian zone: application of WorldView-2 and Landsat 8. PhD thesis. Linköping: Linköping University. Sweden.

Leff B, Ramankutty N, Foley JA. 2004. Geographic distribution of major crops across the world. *Global Biogeochemical Cycles* 18:GB 1009.

Lobell DB, Field CB. 2008. Estimation of the carbon dioxide (CO_2) fertilization effect using growth rate anomalies of CO_2 and crop yields since 1961. *Global Change Biology* 14:39–45.

Lobell DB, Cassman KG, Field CB. 2009. Crop yield gaps: their importance, magnitudes and causes. *Annual Review of Environment and Resources* 34:179–204; doi:10.1146/annurev.environ.041008.093740.

Mander Ü, Helming K, Wiggering H. 2007. Multifunctional land use: meeting future demands for landscape goods and services. *In*: Mander Ü, Wiggering H, Helming K (eds). *Multifunctional Land Use*. Springer. Berlin, Heidelberg. P 1–13.

Mattsson E, Ostwald M, Nissanka S. 2018. What is good about Sri Lankan homegardens with regards to food security? A synthesis of the current scientific knowledge of a multifunctional land-use system. *Agroforestry Systems* 92:1469–1484.

Mattsson E, Ostwald M, Nissanka SP, Pushpakumara DKNG. 2014. Quantification of carbon stock and tree diversity of homegardens in a dry zone area of Moneragala district, Sri Lanka. *Agroforestry Systems*. doi:10.1007/s10457-014-9780-8.

McKay D. 2005. Reading remittance landscapes: female migration and agricultural transition in the Philippines. *Geografisk Tidsskrift-Danish Journal of Geography* 105(1):89–99.

Mead R, Willey R. 1980. The concept of land equivalent ratio and advantages in yields from intercropping. *Experimental Agriculture* 16:217–228.

Meijer S, Catacutan D, Ajayi O, Sileshi G, Nieuwenhuis M. 2014. The role of knowledge, attitudes and perceptions in the uptake of agricultural and agroforestry innovations among small holder farmers in sub-Saharan Africa. *International Journal of Agricultural Sustainability* 13:40–54.

Mohapatra S, Ratha D. 2011. Remittance markets in Africa. The International Bank for Reconstruction and Development/The World Bank. Washington, DC. 380p. http://siteresources.worldbank.org/EXTDECPROSPECTS/Resources/476882-1157133580628/RMA_FullReport.pdf.

Mosquera-Losada MR, Santiago Freijanes JJ, Pisanelli A, Rois M, Smith J, den Herder M, Moreno G, Malignier N, Mirazo JR, Lamersdorf N *et al.* 2016. Extent and success of current policy measures to promote agroforestry across Europe. Deliverable 8.23 for EU FP7 Research Project: AGFORWARD 613520. 95p. www.agforward.eu/index.php/en/extent-and-success-of-current-policy-measures-to-promote-agroforestry-across-europe.html.

Mulia R, Simelton E, Le T, Pham T, Do T. 2018. Native and endangered timber tree species in homegardens of northeast and northcentral Vietnam. *Biodiversity International Journal* 2(2):00041.

Nair P. 1993. An introduction to agroforestry. Kluwer Academic Publisher. Dordrecht. www.worldagroforestry.org/units/library/books/PDFs/32_An_introduction_to_agroforestry.pdf?n=161.

Newton P, Gomez A, Jung S, Kelly T, Mendes T, Rasmussen L, dos Reis J, Rodrigues R, Tipper R, van der Horst D *et al.* 2016. Overcoming barriers to low carbon agriculture and forest restoration in Brazil: the Rural Sustentável project. *World Development Perspectives* 4:5–7.

[OCED] Organisation for Economic Co-operation and Development. 2001. Multifunctionality towards an analytical framework. Organisation for Economic Co-operation and Development. Paris. www.oecd.org/tad/agricultural-policies/40782727.pdf.

Phelps J, Carrasco L, Webb E, Koh LP, Pascual U. 2013. Agricultural intensification escalates future conservation costs. *Proceedings of the National Academy of Sciences of the United States of America* 110(19):7601–7606.

Pigford A-A, Hickey G, Klerkx L. 2018. Beyond agricultural innovation systems? Exploring an agricultural innovation ecosystems approach for niche design and development in sustainability transitions. *Agricultural Systems* 164:116–121.

Ragasa C, Mazunda J. 2018. The impact of agricultural extension services in the context of a heavily subsidized input system: the case of Malawi. *World Development* 105:25–47.

Rahn E, Vaast P, Läderach P, van Asten P, Jassogne L, Ghazoul J. 2018. Exploring adaptation strategies of coffee production to climate change using a process-based model. *Ecological Modelling* 371(C):76–89.

Rosenstock T, Wilkes A, Jallo C, Namoi N, Bulusu M, Suber M, Bernard F, Mboi D. 2018. Making trees count: measurement, reporting and verification of agroforestry under the UNFCCC. CGIAR Research Program on Climate Change, Agriculture and Food Security (CCAFS). Wageningen, the Netherlands. https://hdl.handle.net/10568/98404.

Simelton E, Dam BV, Catacutan D. 2015. Trees and agroforestry for coping with extreme weather events – experiences from northern and central Viet Nam. *Agroforestry Systems* 89(6):1065–1082.

Smith LC, Subandaro A. 2007. Measuring food security using household expenditure surveys. IFPRI International Food Policy Research Institute. Washington, DC. 147p. www.ifpri.org/publication/measuring-food-security-using-household-expenditure-surveys.

Taha E, El-Mahdy A. 2014. Land Equivalent Ratio as a reference for relative crowding coefficient and aggressivity of intercropped plant species. *Middle East Journal of Agriculture Research* 3(3):576–585.

Tschirley D, Reardon T, Dolislager M, Snyder J. 2015. The rise of a middle class in East and Southern Africa: implications for food system transformation. *Journal of International Development* 27:628–646.

[USDA] United States Department of Agriculture. 2015. USDA Coexistence Fact Sheets Conventional Farming. February 2015. Washington, DC. 2p. www.usda.gov/sites/default/files/documents/coexistence-conventional-farming-factsheet.pdf.

Vereijken P. 2003. Transition to multifunctional land use and agriculture. NJAS Wageningen. *Journal of Life Sciences* 50(2):171–179.

Vreeker R. 2004. Urban multifunctional land use and externalities. European Regional Science Association. https://ideas.repec.org/p/wiw/wiwrsa/ersa04p346.html.

[WB] World Bank. 2019. Gini index (online). World Bank Open Data. Development Research Group. Washington, DC. https://data.worldbank.org/indicator/SI.POV.GINI?locations=ET-ZA-NA-BW-KE-NG-BF&name_desc=false&view=chart.

Weltzien E, Christinck A. 2017. Participatory breeding: developing improved and relevant crop varieties with farmers. *In*: Snapp S, Pound B (eds). *Agricultural Systems Agroecology and Rural Innovations for Development*. Academic Press. Cambridge, USA. P 259–301.

Wiggering H, Dalchow C, Glemnitz M, Helming K, Müller K, Schultz A, Stachow U, Zander P. 2006. Indicators for multifunctional land use – linking socio-economic requirements with landscape potentials. *Ecological Indicators* 6(1):238–249.

Wiggering H, Weißhuhn P, Burkhard B. 2016. Agrosystem services: an additional terminology to better understand ecosystem services delivered by agriculture. *Landscape Online* 9:1–15.

Wilson G. 2008. From 'weak' to 'strong' multifunctionality: conceptualising farm-level multifunctional transitional pathways. *Journal of Rural Studies* 24:367–383.

Woodhouse P. 2003. African enclosures: a default mode of development. *World Development* 31:1705–1720.

Zomer R, Neufeldt H, Xu J, Ahrends A, Bossio D, Trabucco A, van Noordwijk M, Wang M. 2016. Global tree cover and biomass carbon on agricultural land: the contribution of agroforestry to global and national carbon budgets. *Nature Scientific Reports* 6. doi:10.1038/srep29987.

2 Nigerian climate-smart agriculture practices with scaling potential

Simon A Shomkegh

Current status of climate-smart agriculture

In Nigeria, agriculture contributes 31 per cent of the GDP but remains the main occupation for more than 70 per cent of the population (SRP 2016). With most of the agriculture being rainfed (NEST and Woodley 2012), production and livelihoods are sensitive to both short-term variations in rainfall patterns and long-term warming (Campbell *et al.* 2011; Jalloh *et al.* 2013). The 1960s experienced a wetter than normal period, while the following two decades were drier than normal (Gommes and Petrassi 1996); this led to famines across the continent. Eleven million Nigerian children are chronically malnourished (UNICEF 2013), one million under age five are affected by severe acute malnutrition (CIFF 2014), and 30 per cent are underweight (NNPC 2013). With rainfall projected to decrease over large parts of Africa (IPCC 2014), animal feed sufficiency will be at risk as the growing periods for crop and fodder will shorten by an estimated 20 per cent on average by 2050 in Western and Southern Africa, causing a 40 per cent decline in cereal yields and biomass (Lobell *et al.* 2011). Without including the humanitarian suffering, in a scenario of no adaptation, climate change is estimated to result in an economic loss of between 2 and 11 per cent of the total gross domestic product by 2020, equivalent to NGN15–69 trillion (US$100–460 billion) (FME 2011). Adapting to these challenges will require changes in agricultural production methods, such as tested climate-smart practices that reduce the risk of crop failures in rainfed agriculture as well as in the consumption patterns, to reduce inefficiencies and waste across the production stages (Victoria *et al.* 2012).

Climate-smart agriculture was coined by FAO in 2010 to address the challenges of ensuring food security for a growing population under the impacts of climate change, while also mitigating greenhouse gas emissions from the agriculture and forestry sectors (FAO 2013). The approach operates on the principles of integrated landscape management and incorporates rural development, biodiversity conservation and ecosystem services (Harvey *et al.* 2014; Scherr *et al.* 2012a). It aims to support the attainment of the Sustainable Development Goals on food security and improved nutrition, combatting climate change and its impacts, supporting sustainable forest management, combatting desertification and halting land degradation and biodiversity loss. Since its launch, climate-smart agriculture has undergone debates regarding the roles and responsibilities of developing countries in reducing global greenhouse gas emissions, and what types of technologies promote sustainable agriculture (Lipper *et al.* 2018; Rosenstock *et al.* 2016, 2019).

In 2011, the CGIAR Research Program on Climate Change, Agriculture and Food Security (CCAFS) and local partners began piloting climate-smart villages in West Africa (Ouédraogo *et al.* 2018). The programme was implemented in Ghana, Mali, Niger, Burkina Faso and Senegal through participatory action research, which is a community process in which farmers and scientists do research and learn together, focusing on changing and reflecting (see, for example, Gonsalves *et al.* 2005). The practices most adopted by farmers in the former three countries were organic manure or compost and integrated farming systems, such as intercropping, because farmers observed improved productivity and more stress-tolerant systems, while in Burkina Faso and Senegal farmers adopted improved crop varieties, soil and water conservation technologies, agroforestry and integrated soil fertility management (Ouédraogo *et al.* 2018). Other practices in the climate-smart village programme were tree planting, agroforestry, early sowing or planting and farmer-managed natural regeneration. Specifically, farmers and scientists identified 20 tree and shrub fodder species that were abundant and had palatable fodder qualities to support livestock farming under periods of weather stress, particularly droughts (Partey *et al.* 2018). The high adoption rates (78 to 90 per cent of farmers) indicate wider scaling potential in the West African region (Ouédraogo *et al.* 2018). Nigeria was not part of the CCAFS programme but had similar research on improved crop varieties, changes in planting dates, zero tillage, natural regeneration, agroforestry, pasture management regimes and rain water harvesting (Cervigni *et al.* 2013). The purpose of this study is to document five common farming practices in the semi-arid Benue State of Nigeria (see Figure 1.5 in Simelton, Ostwald and Osiru Chapter 1) that count as climate-smart agriculture (FAO 2013).

Climate-smart agriculture in Benue

Study site

Benue State in central Nigeria is located between latitudes 6.5° and 8°N and longitudes 6.5° and 10°E, in the southern Guinea savanna ecological zone. The state has a population of 4,250,000 and covers approximately 34,000 square kilometres, of which about 60 per cent is cultivated with crops and 2 per cent is forested. The vegetation consists of dense tall

24 *Simon A Shomkegh*

grasses, riparian forests along rivers, and grasslands with dispersed trees. Rainfall is characterized by the variable onset and cessation of rainy seasons, with increasingly delayed onsets (Adamgbe and Ujoh 2013). For example, between 1980 and 2009, the earliest and latest onset of the rainy season varied by two-and-half months, 20 February (1982) and 5 May (1983), while rainfall cessation varied by two-and-a-half months, 28 September (1983) and 12 December (1988) (Figure 2.1). Between 1960 and 2004, there were two periods with higher annual total rainfall, 1961 to 1969 and 1995 to 1999 (Ologunorisa and Tersoo 2006), peaking at 1757 millimetres in 1963, while the two driest years received half that amount, 841 millimetres in 1988 and 882 millimetres in 1973 (Atedhor 2016).

In the recent decade, 2010 to 2018, the total rainfall amount was normal while the annual seasonal rain period was delayed and both the onset and cessation time frames were later and shorter, with the onset between 9 March and 20 April and cessation between 10 November and 23 December (Figure 2.1). Between onset and cessation, prolonged dry spells of more than ten days, which may affect the maturity of some crops, occurred (SRP).

Farming in the Benue region relies on rainfed cultivation of arable crops (including maize, guinea corn, rice, millet, sesame, soybean, groundnut, cowpea, yam, cassava, potato, vegetables) and perennial tree crops such as citrus, mango, avocado, pear and cashew. Domestic farm animals include pigs, poultry, goats, sheep, cattle and fish. Like any traditional farming

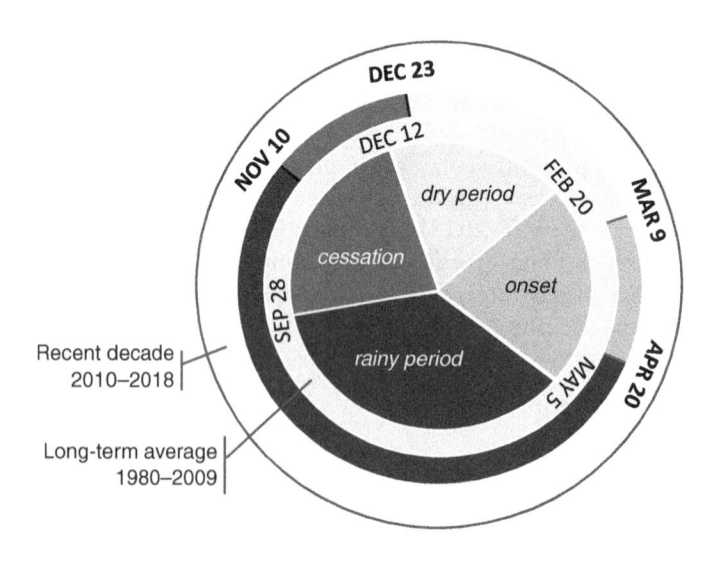

Figure 2.1 Onset and cessation of rainy season in Benue State.
Sources: Adapted from Atedhor 2016 and SRP 2017.

culture, Benue farmers developed 'climate-smart' practices long before the concept was launched, in order to cope with variable rainfall. For example, keeping scattered trees in farmlands, intercropping, and growing orchards have been practised for decades, while zero tillage and improved crop varieties were introduced by the agriculture extension service more recently to strengthen production and income in response to the changing rainfall regimes.

Method

This chapter consists of a review of the literature on five climate-smart agriculture practices in the Benue region and key informant interviews with representatives of relevant government agencies, including the Bank of Agriculture, the Benue Agricultural Development Agency and the Federal Ministry of Agriculture, Makurdi field office, following Cacho *et al.* (2018). For the two most prevalent climate-smart practices, semi-structured interviews with farmers were also conducted in 2018 to elicit information about input costs and benefits of the practices. This included 120 citrus orchard farmers in 12 villages in the Ushongo local government area and 100 farmers practising zero tillage in ten villages in the Guma local government area (see socioeconomic characteristics in Table 2.1). The respondents were randomly selected through consensus among households or individuals who owned their croplands either by inheritance or acquisition. For citrus orchards, some experience and good maintenance practices such as regular weeding, mulching and pruning were additional conditions. Land ownership was not a criteria per se, however, perennial land uses such as orchards or tree plantations are long-term investments, land with insecure tenure, such as a rental agreement, is only used for annual crops. For zero tillage for arable crops, some indigent farmers operate their own lands, while farmers from outside the community may rent land for cultivation. The interviews were conducted with the head of household, who according to customary norms is a man, except for widowed heads of households. The survey covered general farm characteristics, management practices, and farm costs and revenues. The profitability analysis estimated the difference between the economic return on production and the total costs for input and labour, per hectare (Momoh *et al.* 1999). For the statistical test, correlation analysis using the Spearman correlation coefficient was performed on a subset of the households, where cases with missing or clearly deviating values were removed as outliers.

To quantify the efficiency of an intercrop (Atabo and Umaru 2015), the indicators 'land equivalent ratio' (see Figure 1.4 in Simelton, Ostwald and Osiru, Chapter 1) and 'Land Equivalent Coefficient' were used. The Land Equivalent Ratio is defined as the relative land area required of a sole crop to produce the same yield as intercropping (Carlson 2008; Mead and Willey 1980; Willey 1985), where a value above one indicates that the

Table 2.1 Characteristics of interviewed citrus orchard and zero tillage farmers in Benue region

Socioeconomic characteristics	Citrus orchard (n = 120)	Zero tillage (n = 100)
	Per cent	Per cent
Gender		
Men:Women	100:0	91:9
Age (years)		
Less than 30:31 to 50:above 51	14:51:35	31:61:8
Marital status		
Single:Married	0:100	11:89
Education (level)		
No formal:Primary:Secondary:Tertiary	0:35:34:28	20:19:50:11
Household size (persons)		
Less than 5:5 to 9:10 to 15:more than 15	23:47:13:17	11:70:9:10
Farm size (hectares)		
Less than 0.5	10	6
0.5 to 1.0	20	38
1.1 to 2.0	31	21
2.1 to 4.0	14	33
More than 4.0	25	2
Source of labour		
Household:Hired:Community members	81:13:6	57:43:0
Membership in farm development association		
No:Yes	80:20	77:23

Source: Author's field data 2018.

productivity per unit area is higher for a crop grown in a mix compared to monoculture. The land equivalent coefficient measures the efficiency of the interaction of the intercrop mix (Ofori and Stern 1987), where a value above 0.25 indicates good efficiency and monocultures have values less than 0.25 as there is no mix to be measured.

Five climate-smart agriculture practices in the Benue region

The five most common climate-smart agriculture practices in the Benue region of central Nigeria include traditional parklands, intercropping, citrus orchards, zero tillage and improved varieties. Here presented in chronological order.

Traditional parklands: indigenous trees on farmlands

For generations, farmers have kept indigenous trees scattered on their farmlands, especially trees with high economic value and open canopies. Aged and unproductive tree species are removed for various uses including fuelwood, crafts-making and charcoal production while tree replacement in the region mainly occurs by natural regeneration during fallow periods. Van Gelder and O'Keefe (1995) found that trees left on farmlands served as household reserves for construction material, medicine and food. In places with abundant trees with dense or nearly closed canopies, farmers thin them by cutting some trees to reduce the competition between trees and crops. Farmers maintain stump regrowth by pruning at weeding and using the prunings as mulch. Stumped trees provide green fodder and support crop growth through leaf litter decomposition while the stumped tree roots enhance soil stability (Bayala *et al.* 2014; Shomkegh *et al.* 2016).

In the Benue region, common tree species kept as standing trees or stumps include ironwood (*Prosopis africana*) (Figure 2.2), African locust bean (*Parkia biglobosa*), shea tree (*Vitellaria paradoxa*), Cape fig (*Ficus sur*), wild custard-apple (*Anona senegalensis*), borassus palm (*Borassus aethiopum*) and acacia (*Acacia nilotica*) (Shomkegh *et al.* 2016). As in Burkina Faso (Sanou Chapter 3), the shea tree can be intercropped with various crops, including millet, sorghum, maize, pigeon pea, cotton, cowpea and cassava (Ani *et al.* 2012; Bayala *et al.* 2014). The fruit pulp is rich in vitamins A and B (Ugese *et al.* 2008), and the fatty kernel is used for production of shea butter (Ani *et al.* 2012). Furthermore, the larva of the pallid emperor moth (*Cirina forda Westwood*), which feeds on shea leaves, contains high levels of protein and potassium (Omotoso, 2006) and is consumed in West Africa, Southern Africa, and the Amazon (Agbidye *et al.* 2009a,b; Amatobi 2007; Mbata and Chidumayo 2003). Farmers with shea trees in the Ukum local government area of Benue State could earn between NGN80,000 (US$222) and NGN1,000,000 (US$2,778) per hectare per year, depending on the tree density, compared to a maximum

Figure 2.2 Ironwood trees in a maize field in Makurdi, Benue State.
Photo credit: Shomkegh 2018.

of NGN60,000 (US$167) for those without shea trees (Ani *et al.* 2012). For many households, the fruits are collected and provide income during the hungry months (Hammond *et al.* 2019).

Intercropping

Intercropping involves two or more crops grown close to each other simultaneously, often in rows (Okpara *et al.* 2005). This strategy to minimize risks for crop failure and optimize land use (Ullah *et al.* 2007; Undie *et al.* 2013) depends on finding the best combination of crops that can compete with monoculture in terms of production and interaction effects for each context (Seran and Brintha 2010). For example, intercropping with legumes that fix nitrogen has been shown to improve soil health and control certain pests and diseases (Nyasimi *et al.* 2014).

Studies in the Benue region (Idoko *et al.* 2018*a-c*; Ijoyah and Dzer 2012; Ijoyah *et al.* 2012) have identified several intercrop combinations that render higher maize yields than monocultures. In particular, maize (*Zea mays* L.) intercropped with okra (*Abelmoschus esculentus* L. *Moench*) showed no significant difference in yield, alone and when intercropped. Here, the intercropped system showed the lowest competitive pressure, as indicated by high land equivalent ratios 1.78 and 1.75 in 2010 and 2011, respectively (Ijoyah and Dzer 2012). In Makurdi, all five improved maize varieties intercropped with sweet potato gave a Land Equivalent Ratio above 1.0 (Idoko *et al.* 2018*a*) and a land equivalent coefficient above 0.25. See also the chapter about maize in intercrops (Adewopo Chapter 7).

Citrus orchards

Citrus orchards have been established on arable cropland, fallow and degraded lands. Technically, the practice evolved in 1986 when the Agricultural Development Programme's extension service system started offering advice and training on crop production, agroforestry, livestock production and fisheries.

Sweet orange varieties were budded onto seedlings of rough lemon (*Citrus jambhiri* Lush). Budding is a propagation method whereby a bud of a plant with desired qualities is joined with another plant for improved productivity. Here, farmers preferred rough lemon as rootstock due to its resistance to diseases and good production (Ortese *et al.* 2012). The lemon was intercropped at a spacing of seven by seven metres with arable crops, such as cassava (Figure 2.3), soybean, cowpea, maize, sweet potato and groundnut, especially in the early years of the citrus orchard. Orchard farmers may also have a few other fruit trees in homesteads, such as mango and papaya, for household consumption.

Maintenance in orchards generally involves raising rough lemon seedlings in the nursery, budding, land preparation, planting of budded seedlings, weeding, fertilizer application, pruning and mulching. Mulching with residues from weeding and pruning is done to improve soil fertility, aeration and percolation around the citrus trees, while burning of organic residuals is restricted to avoid fire damaging the trees. In Kenya, this practice is reported to recycle organic nutrients, sequester carbon and provide year-round ground cover and retention of organic matter and water in the soil (Scherr *et al.* 2012*b*). The transition from arable to tree crops is estimated to increase soil carbon by 50 to 100 per cent (Glover and Reganold 2010) over a four-year period. Where legumes were included to improve nitrogen fixation (Onoja Chapter 4), fertilizer consumption decreased.

In this study, the orchards are small family ventures. More than half of the surveyed orchards were grown on land owned by the family (55 per

Figure 2.3 A citrus orchard intercropped with cassava in the Benue region.
Photo credit: Shomkegh 2018.

cent) and dependent on family labour (81 per cent) (Table 2.1). Also, most orchards were on plots smaller than two hectares (61 per cent), while 25 per cent of the interviewees had larger farms, more than four hectares. Compared to zero tillage, the orchard farms covered more diverse socio-economic characteristics. This may depend on it taking a longer time to establish an orchard, and land and practices having stayed in the family, passed on through inheritance (Ortese *et al.* 2012). The average cost for labour and inputs was NGN30,000 per hectare, resulting in a profitability ratio with a gross income up to 25 to 45 times the cost, depending on farm gate prices and farm size. Here, the selling price of orange was calculated for three rates at NGN1000, 1,500 or 1,900 per bag of 50 kilogrammes, and only results for the lowest price are shown in Figure 2.4. Incomes from

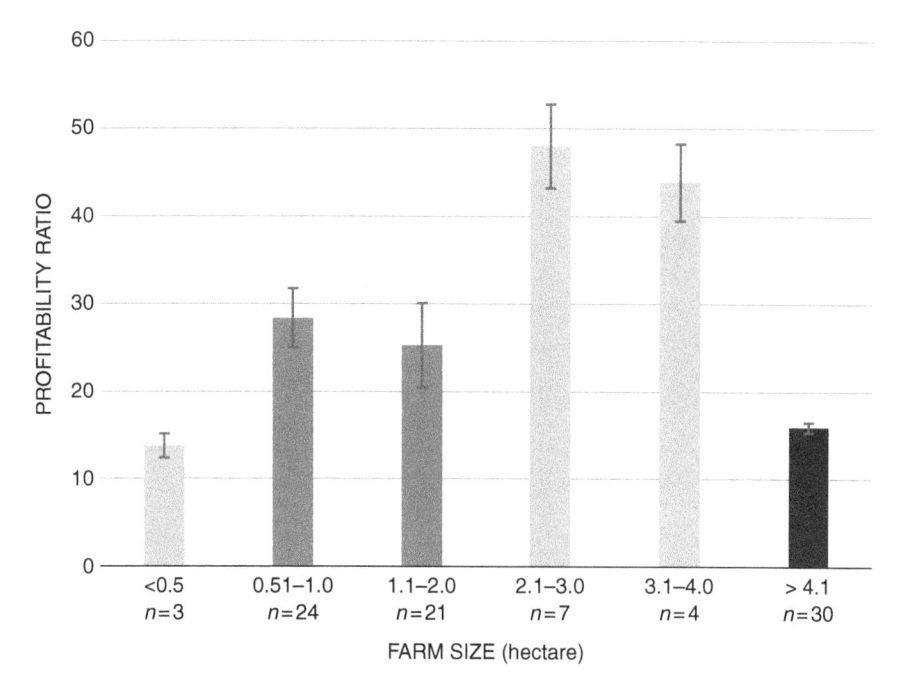

Figure 2.4 Profitability ratio, gross income to cost, from citrus orchards per farm size ($n=92$). Error bars denote standard error of mean. The shade of the bars represent the number of respondents (n) per farm size category.

Source: Author's field data 2018.

other crops in the orchard are highly variable and excluded from the analysis. The profitability ratio per hectare was similar for farms between one-half and two hectares and the lowest for the largest farms, and was strongly correlated with higher education, larger household size, source of household labour and farm income (Table 2.2). This may partly be explained by more data being available and gave the strongest results for farm sizes between 0.5 and two hectares and above four hectares, while most of the missing data was among the smallest farm sizes. Other explanations could be that larger farms depend on hired labour or are short of labour, or that their citrus trees are more sparsely planted. While some citrus varieties may produce two harvests annually, harvesting for commercial purposes is only done once per year. Some 17 per cent of the orchard farmers also had honey production.

The top three challenges mentioned by the farmers related to infrastructure and access to road networks to reach markets, high input costs, drought and other climatic stress factors. Orchards are more input-demanding than zero tillage, and although herbicide use was uncorrelated

Table 2.2 Correlation coefficients for profitability ratio per hectare versus farmer characteristics for orchard and zero tillage (see Table 2.1)

Demographic characteristics	CSA practice	
	Orchard (n = 89)	Zero tillage (n = 88)
Age	0.139	–0.006
Education [no formal, primary, secondary, tertiary]	–0.238*	–0.021
Household size	–0.250*	–0.184
Farm size	–0.146	–0.490**
Type of land ownership [individual, family, rented]	–0.163	–0.274**
Labour [household, hired, mixed]	–0.579**	–0.148
Farm income	0.256*	–0.206

Source: Author's field data 2018.

Notes
Spearman rank correlation (2-tailed) ** significant at the 0.01 level; * significant at the 0.05 level.

with education levels, misuse was observed. Half of the interviewed farmers said they needed training, especially on fertilizer use, and pest and disease control, while the other half said they needed no training.

Zero tillage

Traditionally, fields were cleared and ploughed with hand hoes before planting. However, with the shorter rainy seasons, planting without tillage was introduced to save cropping time, maintain soil structure and prevent soil erosion (Figure 2.5). With zero or minimum tillage, soil and surface residues are minimally disturbed (Parr et al. 1990), and the need for manual and mechanical seedbed preparation before planting is eliminated (Lal 1983). Minimum tillage with cover crops and mulch can enhance the soil organic matter, while also supporting biological processes and nutrient and hydrological cycling (Hobbs and Govaerts 2009; Milder et al. 2011).

Zero tillage has become a common practice in the Guma local government area for crops such as melon, maize, millet, rice and cowpea. One-third of the surveyed farmers began this practice in 2008 and observed immediate benefits, particularly with respect to conserving soil moisture and, more importantly, saving time and labour costs. Prior to zero tillage, crop rotations with tubers that require tillage may have been practised, including yam, cassava and coco yam. Here, zero tillage starts with herbicide application on selected plots; plant residuals are left to decay on the soil surface to improve the soil organic matter. Although plots are only about one hectare (Table 2.1), preparations for sowing do need to be fast

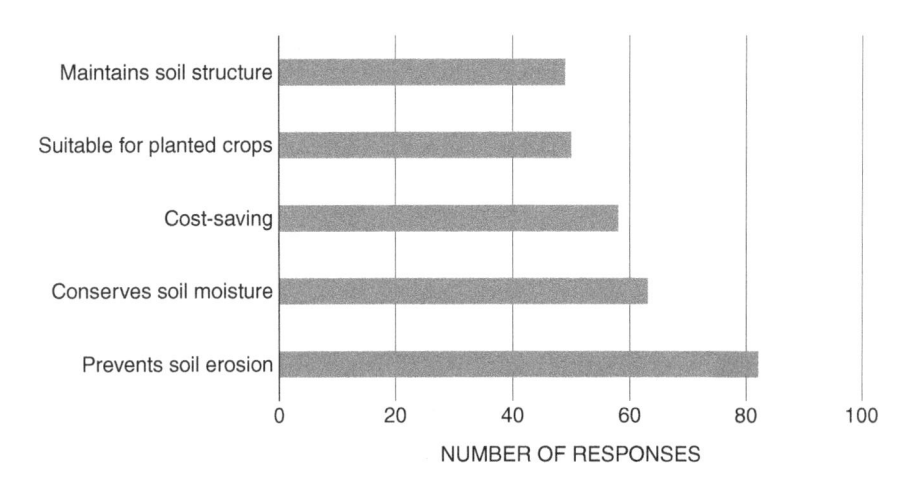

Figure 2.5 Farmers' reasons for adopting zero tillage in the Benue region, per cent of respondents (*n* = 100, multiple responses were possible).

Source: Author's field data 2018.

to benefit from the rains. Herbicides are effective also when diluted and are relatively affordable even for the poor; farmers prefer them to manual weeding to save time and reduce soil disturbance in the event of surface run-off. Next, crops are planted, often improved varieties. Although different crops are grown using zero tillage, nearly all are harvested only once per year. Farmers did not report spread of diseases with this practice.

The average cost for labour and inputs was considerably lower than for orchards, about NGN20,000 per hectare, and the gross income from selling the crops was NGN275,000 – thus a considerable profitability. Both the costs and the income from sales per hectare were the lowest for farms smaller than two hectares. However, the profitability was significantly higher for farms below one hectare and non-rented farm land (Figure 2.6, Table 2.2). Here, 12 per cent of the data points were removed as missing data and outliers. Moreover, most farmers practising zero tillage were comparatively young, aged between 21 and 40 years old (65 per cent), which according to Halima and Edoja (2013) could help explain their greater returns. In this study, unmarried men had significantly higher net benefits as the household consumption was lower than for families. On the other hand, the younger households may face labour shortage, explaining the use of herbicides.

Improved varieties: drought and disease resistant crops

Generally, 80 per cent of cowpea yields are lost due to the parasite weed *Striga gesnerioides*, whose outgrowths (haustoria) penetrate the host root

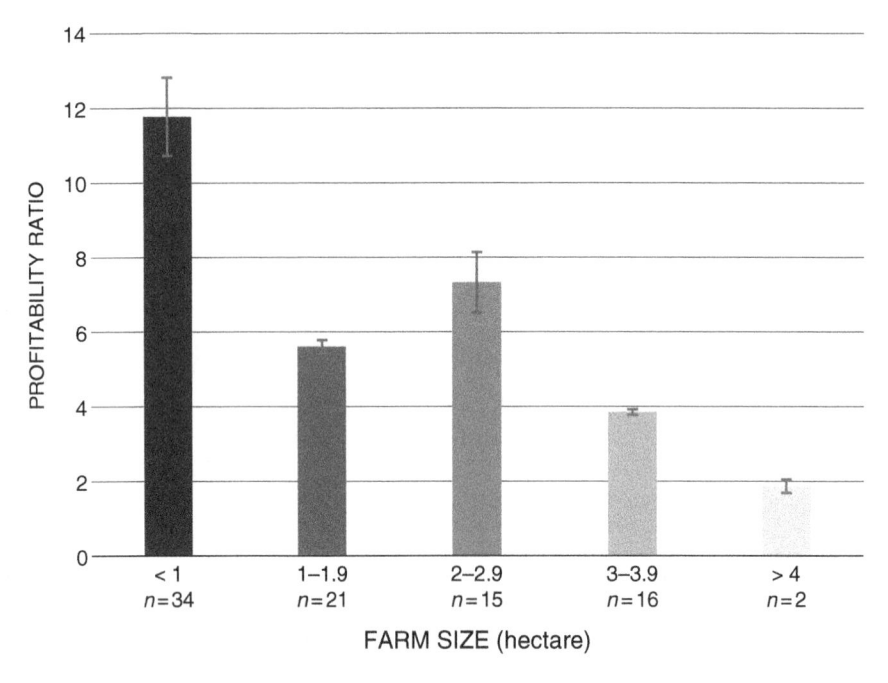

Figure 2.6 Profitability ratio, gross income to costs, for zero tillage by farm size (*n* = 88). Error bars denote standard error of mean. The shade of the bars represent the number of respondents (n) per farm size category.

Source: Author's field data 2018.

and absorb nutrients (Omoigui *et al.* 2017). Furthermore, in response to the increasing variability of the rainy season, scientists from universities and research institutes are developing improved crop varieties with shorter maturity periods and better resistance to pests and diseases. Since 1987, when the National Centre for Genetic Resources and Biotechnology was established, new crop varieties have been registered and released to farmers every year. For example, in 2017, 595 high-yielding drought, disease and pest resistant varieties were released and catalogued, ranging from tubers, cereals and forage legumes to vegetables (NACGRAB 2016). Among those were two drought-tolerant varieties of cowpea (FUAMPEA 1 and FUAMPEA 2) from the Federal University of Agriculture in Makurdi, which produced about two tonnes per hectare in the experimental fields and showed strong resistance to the parasitic weeds *Striga gesnerioides* and *Alectra vogelii* (Omoigui *et al.* 2017). In 2017, for the first time, Benue farmers were able to cultivate early maturing cowpea without competition from the parasitic plants or loss due to the variable rainfall.

Factors for success in studied cases

The reviewed literature and two surveys indicate that the five climate-smart agriculture practices presented here are all low-cost changes that can lead to more stable or higher yields and incomes, compared to monocultures. Also, these practices have been proven resilient to the variable rainfall patterns in the region. Common factors that contributed to the success include the following.

Meeting farmers needs

The new varieties delivered what farmers needed. The drought and disease resistant crop varieties quickly replaced traditional varieties because the former offered a higher yield, matured faster and reduced the amounts of water and nutrients required. The zero-tillage practice saved tillage costs and contributed to soil improvement. Citrus orchards and intercropping produced more with more efficient land°use.

Committed research

Collaboration between local research institutions and farmers ensured continuous improvement of crop varieties that met farmers' demands. Varieties that matured earlier, tolerated drought better and were resistant to pests and disease addressed the challenges posed by both variable rainfall and food security needs. Every year, the National Centre for Genetic Resources and Biotechnology approves varietal registration and released proposals from different research institutions across the country and keeps informing about new varieties.

Policies and credit

SMS service to farmers' cell phones was used in communicating input availability and distribution in nearby centres. The federal government provided subsidized inputs for priority crops, such as organic and inorganic fertilizers, improved seeds, and micro-nutrients. This removed the price escalation and delay in input availability. The Central Bank of Nigeria provides loans at 9 per cent interest rate per year through the Anchor Borrowers Scheme, compared to 22 per cent in commercial banks.

Some farmers received grants from the International Fund for Agriculture and Development and the World Bank for inputs and technologies for the federal government's priority crops. The loans were accessed through commodity cooperative groups of ten to 15 farmers who must be registered with relevant government agencies and supervised by the funding agency. By the end of 2017, the private sector Anchor Borrowers Scheme had provided loans to 1,758 soybean farmers in Benue State, while the government scheme had reached 9,096 farmers. Periodically, the scheme supported training in crop performance monitoring. Furthermore, recognized value-chain-enhancing organizations known as off-takers bought the harvests to ensure optimum value for produce for farmers in the scheme.

Limitations

For the further expansion of climate-smart practices, a few critical bottlenecks need to be resolved.

Insufficient harvest and processing technologies

Affordable and feasible harvesting and processing technologies for fruits and staple crops, such as cereals and tubers, are unavailable in the region. This currently limits the scaling potential and has led to food waste and post-harvest income losses for crops that perish shortly after harvest. For example, tomato, yam and citrus are harvested in large quantities over particular periods, which pushes down the price for producers due to oversupply at peak harvest periods. Investments from the government, donor organizations and the private sector in affordable harvesting and processing technologies can be targeted to reduce harvest and post-harvest losses as well as enable higher and more stable incomes for rural farmers.

Expensive agricultural inputs

Expenses for agricultural inputs, such as improved seeds, herbicides and fertilizers are high and disadvantage those farmers who are not yet covered by national or non-governmental organizations' support schemes, both financially and by reducing their opportunity to effectively respond to rainfall variability. Targeted public investment in the agricultural sector is needed to ensure support is available to all rural farmers. This investment will pay back in avoided crop losses and contribute to ensuring national food security and thus to attaining the Sustainable Development Goals.

Unregulated use of herbicides on croplands

For those who can afford it, increased use of herbicides has replaced manual weeding, as weeding is considered time-consuming compared to chemical methods (Shomkegh *et al.* 2012). Persistent use of herbicides may lead to environmental and food safety risks. Further, the abundance of some grass species is declining, which could lead to loss of biodiversity. For example, spear grass (*Heteropogon contortus* (L.) Beauv. Ex Roem. & Schult), widely used in the region for thatching local houses, is becoming scarce and is leading to less suitable thatching alternatives from crop residues such as soybean stalks. Awareness-raising activities for extension workers, farmers, agriculture service providers, and policy makers will be important for plant protection and to optimize agrochemical application use and techniques, conducting periodic soil tests and promoting traditional weeding methods such as using holes and cutlasses.

Fading agricultural extension services

The agricultural extension service system in Nigeria is generally underfunded and understaffed. According to key informant interviews, the extension service system in Benue State began in 1986 with 400 staff (300 men and 100 women), with each extension worker covering 1,500 farmers. The female extension workers demonstrated specific technologies targeting women, such as soybean processing, vegetable preservation, and root and tuber processing and storage. In 2004, the state only had 211 extension agents who made 20,497 farm visits advising on crop production, livestock, fisheries and agroforestry (BNARDA 2004). Currently, the number of extension workers has decreased to 32, due to retirement and the lack of recruits. Consequently, the remaining extension workers have more farmers to support, about 5,000 for each worker, which can be compared to the national average of 1,200 farmers (FTF 2016). One proposed indicator for the Sustainable Development Goal of ending hunger is 'Number of agricultural extension workers per 1,000 farmers' (SDSN 2015), with increased investment in extension services as a relevant target, but so far there is no guidance. Effective extension service workers also need to be trained and updated on agronomic practices, interpreting climate information, and alternative climate-smart farming methods to sustainably improve farm productivity, with public investment in university courses, extension services and rural infrastructure (FTF 2016).

Cumbersome land acquisition process

Land administration falls under the Land Use Act of 1978. However, customary practices prevail. The Land Use Act complicates the process of securing and perfecting land titles for agricultural production to the extent that about 95 per cent of agricultural lands are untitled (APP 2016). Land tenure and land acquisition processes are cumbersome, time-consuming and costly, which makes it difficult for farmers to obtain land titles for agricultural production. This prevents farmers from using their land as collateral for long-term investments and access to commercial loans. Rural women are particularly affected because, traditionally, only men are entitled to own land, as heads of households (see also Sanou Chapter 3 and Onoja Chapter 4). Women may be granted access to cultivate arable crops and, sometimes, intercrops in orchards, but perennial tree plants are owned and inherited through paternal lines. According to Kasimbazi (2017), secure land and property rights are critical for reducing poverty and for enhancing economic development, gender equality, social stability and sustainable resource use. A review of laws related to tenure is needed at the federal and state government levels both to relax the land acquisition process for farmers so that women can have equal opportunities to own land and to support large-scale long-term agricultural investments.

Policy aspects

Although Nigeria has no direct strategy for climate-smart agriculture and the Land Use Act presents an obstacle, the following five government policies at the state or federal level have the potential to support more widespread adoption.

National Adaptation Strategy and Plan of Action for Climate Change

The National Adaptation Strategy and Plan of Action for Climate Change seeks to minimize climate risks, improve local and national adaptive capacity, and leverage new opportunities for facilitating international collaboration (FME 2011). The policy supports improved agricultural systems and practices for crops and livestock and access to climate information, such as early warning and meteorological forecasts, with stated roles and responsibilities of the federal, state and local governments, the private sector, civil society organizations, communities and individuals for these improved systems and practices. The policy also emphasizes the link between improved management of natural resources and climate adaptation actions in agriculture. If all stakeholders commit to meeting their responsibilities, this will increase the impact of the policy on climate-smart agriculture practices in the region.

Intended nationally determined contribution

Nigeria's third intended nationally determined contribution recognized that climate-smart agriculture is a key means towards meeting the ambitions of agricultural transformation. The document (FGN 2015), which was submitted to the 2015 United Nations Framework Convention on Climate Change, aims to sustainably increase agricultural productivity and support equitable increases in farm incomes, enhancing food security and development while reducing greenhouse gas emissions. The recommended practices include halting deforestation and promoting agroforestry. The estimated benefits from agroforestry include total (lifetime) carbon emission reductions ranging between 158 million tonnes and 712 million tonnes (FGN 2015).

Agricultural Promotion Policy supports climate-smart agriculture

The Agricultural Promotion Policy (2016–2020) evolved through an inclusive stakeholder consultation process among farmer groups, academia and private sector investors and has climate-smart agriculture as one of its thematic approaches to agricultural development (APP 2016). The policy promotes effective management of natural resources and best practices for climate change adaptation and mitigation that can lead to sustainable agricultural productivity for national food security; however it requires more investment to reach the broader farming population.

The Growth Enhancement Scheme supports inputs

The Growth Enhancement Scheme is driven by the federal government to promote biometric registration of farmers and provide targeted means-based input subsidies (ATA 2011). This policy aims to address the problems with delayed distribution of subsidized inputs and of intermediaries who overcharge farmers and delay the distribution process. The scheme ensures timely delivery of farm inputs directly to farmers via the use of mobile phones. Messages specify types and quantities of farm inputs provided, their subsidized cost and collection centres with a bar code. Farmers then proceed to the centre for verification, payment and pick up the subsidized inputs. This provides improved access to basic farm inputs for registered farmers. By 2017, over 295,000 farmers were registered in the 23 local government areas of Benue State (FMARD 2017). Strategies for how to sustain and scale the scheme are outlined in FTF (2016).

Prohibition of open grazing and Ranches Establishment Law

Benue State's Open Grazing Prohibition and Ranches Establishment Law was enacted in 2017 to resolve conflicts between herders and farmers in their struggle for scarce natural resources. Partly related to the weak tenure systems, the competition for fertile arable lands and grazing areas deepened with the increasing population, urbanization and climatic stress. Moreover, some argue that the trampling by cattle distorts wildlife habitats, compacts the soil and leads to increased run-off. The law took effect in December 2017 and prohibits open grazing of livestock and promotes ranches. As such, courts now have a legal instrument for prosecuting violators. This is intended to lead to peace between herders and farmers and to enhance sustainable crop and animal productivity. The herders are now expected to cut grass or buy fodder and reduce conflicts with farmers over destroyed crops and property. The smooth implementation of the law is challenged by resistance from herders who require capital to acquire land, secure it and obtain pasture for their animals.

Lessons learned from the case

- Changing rainfall patterns are already challenging farmers to identify new adaptation strategies.
- Mixed fruit orchards and the retention of indigenous tree species on farmlands promote landscape restoration and tree cover and provide food and income to farmers.
- Development and adoption of improved drought and/or disease resistant crop varieties for rainfed crop cultivation enhances crop productivity in the context of a changing climate.
- Zero tillage enables farmers to adjust farming calendars in response to rainfall patterns, stabilizes and retains carbon in soils, and reduces farming costs.
- Intercropping returned both higher crop yields through more efficient land use and agro-biodiversity benefits.
- Some state and national policies support climate-smart agriculture, while tenure is ambiguous.
- Additional investments in the extension system are needed to spread good farming practices.

References

Adamgbe EM, Ujoh F. 2013. Effect of variability in rainfall characteristics on maize yield in Gboko, Nigeria. *Journal of Environmental Protection* (4):881–887.

Agbidye FS, Ofuya TI, Akindele SO. 2009a. Marketability and nutritional qualities of edible forest insects in Benue State, Nigeria. *Pakistan Journal of Nutrition* 8(7):917–922.

Agbidye FS, Ofuya TI, Akindele SO. 2009b. Some edible insect species consumed by the people of Benue State, Nigeria. *Pakistan Journal of Nutrition* 8(7):946–950.

Amatobi CI. 2007. *Arthropod Pests of Crops in Nigeria: General Biology, Natural Enemies and Control*. PA Ndahi Printing. Zaria, Nigeria. 252p.

Ani DP, Aondona G, Soom A. 2012. Economic analysis of shea butter plant in Ukum local government, Benue State, Nigeria. *American-Eurasian Journal of Agronomy* 5(1):10–18. doi:10.5829/idosi.aeja.2012.5.1.2032.

[APP] Agricultural Promotion Policy. 2016. The agricultural promotion policy of the Federal Government of Nigeria 2016–2020. *Federal Ministry of Agriculture*, Abuja. 59p. http://fscluster.org/sites/default/files/documents/2016-nigeria-agric-sector-policy-roadmap_june-15-2016_final1.pdf.

[ATA] Agricultural Transformation Agenda. 2011. The agricultural transformation policy of the Federal Government of Nigeria 2011–2016. *Federal Ministry of Agriculture*, Abuja. 59p. http://fscluster.org/sites/default/files/documents/2018-nigeria-agric-sector-policy-roadmap_june-15-2016_final1.pdf.

Atabo JA, Umaru TM. 2015. Assessing the land equivalent ratio (LER) and stability of yield of two cultivars of sorghum (*Sorghum bicolor* L. Moench)-Soyabean (Glycine max L. Merr) to row intercropping system. *Journal of Biology, Agriculture and Healthcare* 5(18):144–149.

Atedhor GO. 2016. Growing season rainfall trends, alterations and drought intensities in the Guinea savanna belt of Nigeria: implications on agriculture. *Journal of Environment and Earth Science* 6(3):1–13.

Bayala J, Sanou J, Teklehaimanot Z, Kalinganire A, Ouédraogo SJ. 2014. Parklands for buffering climate risk and sustaining agricultural production in the Sahel of West Africa. *Current Opinion in Environmental Sustainability* 6:28–34.

[BNARDA] Benue State Agricultural and Rural Development Authority. 2004. Annual Report 2004. Benue Agricultural Development Project. Prepared by the Planning, Monitoring and Evaluation Division. Makurdi, Benue State. 27p.

Cacho O, Paolantonio A, Branca G, Cavatassi R, Arslan A, Lipper L. 2018. Identifying strategies to enhance the resilience of smallholder farming systems: evidence from Zambia. *In*: Lipper L, McCarthy N, Zilberman D, Asfaw S, Branca G (eds). *Climate smart agriculture: building resilience to climate change.* FAO Natural Resource Management Policy 52. Cham, Switzerland. P 425–441. www.fao.org/3/a-i7931e.pdf.

Campbell B, Mann W, Meléndez-Ortiz R, Streck C, Tennigkeit T. 2011. Addressing Agriculture in Climate Change Negotiations: A Scoping Report. Washington, DC. Meridian Institute https://hdl.handle.net/10568/10306.

Carlson JD. 2008. Intercropping with maize in sub-arid regions. Community planning and analysis. Technical brief. Michigan Technical University, Houghton MI. P 2–5.

Cervigni R, Riccardo V, Monia S. (eds). 2013. Toward climate-resilient development in Nigeria. Directions in development. World Bank. Washington, DC. 215p. doi:10.1596/978–0-8213–9923–1.

[CIFF] Children's Investment Fund Foundation. 2014. Community Management of Severe Acute Malnutrition, Nigeria. London UK. https://ciff.org/priorities/childhood-adolescence/nutrition/.

[FAO] Food and Agriculture Organization of the United Nations. 2010. *Opportunities and Challenges for a Converging Agenda: Country Examples.* (FN24; World Bank report prepared for the 2010 The Hague Conference on Agriculture, Food Security and Climate Change). World Bank. Washington, DC.

[FAO] Food and Agriculture Organization of the United Nations. 2013. Climate smart agriculture sourcebook. Rome. 570p. www.fao.org/3/a-i3325e.pdf.

[FGN] Federal Government of Nigeria. 2015. Nigeria's Intended Nationally Determined Contribution. Federal Ministry of Environment. Abuja. www4.unfccc.int/sites/ndcstaging/PublishedDocuments/Nigeria%20First/Approved%20Nigeria's%20INDC_271115.pdf.

[FMARD] Federal Ministry of Agriculture and Rural Development. 2017. FMARD Makurdi Greenhouse Update, Benue state, Nigeria. Report from Prroject Office. Makurdi, Benue State.

[FME] Federal Ministry of Environment. 2011. National Adaptation Strategy and Plan of Action on Climate Change for Nigeria (NASPA-CCN). Building Nigeria's Response to Climate Change (BNRCC) Project Ibadan, Oyo State, Nigeria. http://csdevnet.org/wp-content/uploads/NATIONAL-ADAPTATION-STRATEGY-AND-PLAN-OF-ACTION.pdf.

[FTF] Feed the Future. 2016. Nigeria: In-depth Assessment of Extension and Advisory Services Developing Local Extension Capacity (DLEC) Project March 2017. The US Government's Global Hunger and Food Security Initiative. DigitalGREEN. 64p. www.digitalgreen.org/wp-content/uploads/2017/09/Nigeria-In-Depth-Assessment.pdf.

Glover JD, Reganold JP. 2010. Perennial grains: food security for the future. *Issues in Science and Technology* 26(2):41–47.

Gommes R, Petrassi F. 1996. Rainfall variability and drought in Sub-Saharan Africa. Extracted from FAO agrometeorology series working paper No 9. P 1–7. www.fao.org/3/a-au042e.pdf.

Gonsalves J, Becker T, Braun A, Campilan D, de Chavez H, Fajber E, Kapiri M, Rivaca-Caminade J, Vernooy R (eds). 2005. Participatory research and development for sustainable agriculture and natural resource management. A Sourcebook Volume 1: understanding participatory research and development. International Potato Center-Users' Perspectives with Agricultural Research and Development, Laguna, Philippines and International Development Research Centre, Ottawa, Canada. 272p. https://idl-bnc-idrc.dspacedirect.org/bitstream/handle/10625/35115/IDL-35115.pdf?sequence=1&isAllowed=y.

Halima CI, Edoja MS. 2013. Exploring the relationship between farming practices and vegetation dynamics in Benue State, Nigeria. *African Journal of Geography and Regional Planning* 3(1):218–225.

Hammond J, van Wijk M, Pagella T, Carpena P, Skirrow T, Dauncey V. 2019. Shea butter: a pro-poor, pro-female route to increased income. *In*: Rosenstock TS, Nowak A, Girvetz E (eds). *The Climate-Smart Agriculture Papers. Investigating the Business of a Productive, Resilient and Low Emission Future*. Springer Open. Wageningen, The Netherlands. Chapter 18. https://link.springer.com/book/10.1007%2F978-3-319-92798-5.

Harvey CA, Chacón M, Donatti CI, Garen E, Hannah L, Andrade A, Bede L, Brown D, Calle A, Chará J, Clement C, Gray E, Hoang MH, Minang P, Rodríguez AM, Seeberg-Elverfeldt C, Semroc B, Shames S, Smukler S, Somarriba E, Torquebiau E, van Etten J, Wollenberg E. 2014. Climate-smart landscapes: opportunities and challenges for integrating adaptation and mitigation in tropical agriculture. *Conservation Letters* 7(2):77–90.

Hobbs PR, Govaerts B. 2009. How conservation agriculture can contribute to buffering climate change. *In*: Wallingford RM (ed.). *Climate Change and Crop Production*. CAB International. Oxford, United Kingdom. P 177–199.

Idoko JA, Akaazua BW, Oga JI. 2018*a*. Evaluation of five improved maize varieties for intercropping with sweet potato in Makurdi, southern guinea savanna ecology of Nigeria. *Asian Research Journal of Agriculture* 8(3):1–11. doi:10.9734/ARJA/2018/40425.

Idoko JA, Iorlamen T, Offordile AE. 2018*b*. Effect of intercropping some crop species with orange flesh sweet potato on the performance of orange flesh sweet potato varieties in Makurdi. *International Journal of Agricultural Policy and Research* 6(3):28–37. doi:10.15739/IJAPR.18.004.

Idoko JA, Iorlamen T, Osang PO. 2018*c*. Influence of cowpea varieties and time of introduction of cowpea into maize cowpea intercropping system in Makurdi, southern Guinea savannah, Nigeria. *International Journal of Agriculture and Environmental Research* 4(2):468–485.

Ijoyah MO, Bwala RI, Iheadindueme CA. 2012. Productivity of cassava-sweet potato intercropping system as influenced by varying lengths of cassava cutting at Makurdi, Nigeria. *Journal of Biology, Agriculture and Healthcare* 2(5):1–7.

Ijoyah MO, Dzer DM. 2012. Yield performance of Okra (*Abelmoschus esculentus* L. Moench) and Maize (*Zea mays* L.) as affected by time of planting maize in

Makurdi, Nigeria. *International Scholarly Research Network* 1(1):1–7. doi:10.5402/2012/485810.

[IPCC] Intergovernmental Panel on Climate Change. 2014. Climate Change: Impacts, Adaptation and Vulnerability. A Contribution of Working Group II to the Fifth Assessment Report of the Intergovernmental Panel on Climate Change. *In*: Field CB, Barros VR, Dokken DJ, Mach KJ, Mastrandrea MD, Bilir TE, Chatterjee M, Ebi KL, Estrada YO, Genova RC, Girma B, Kissel ES, Levy AN, MacCracken S, Mastrandrea PR, White LL. (eds.). Cambridge University Press, Cambridge, United Kingdom, P 1–32.

Jalloh A, Nelson GC, Thomas TS, Zougmoré RB, Roy-Macauley H. 2013. West African agriculture and climate change: a comprehensive analysis. International Food Policy Research Institute (IFPRI), Washington, DC. 444p. www.ifpri.org/publication/west-african-agriculture-and-climate-change.

Kasimbazi E. 2017. Land tenure and rights for improved land management and sustainable development. *Global Land Outlook Working Paper*. United Nations Convention to Combat Desertification, Bonn. P 1–28. https://knowledge.unccd.int/publications/land-tenure-and-rights-improved-land-management-and-sustainable-development.

Lal R. 1983. No-till farming: soil and water conservation and management in the humid and sub-humid tropics. International Institute of Tropical Agriculture (IITA). Monograph 2. IITA Ibadan, Nigeria. https://library.wur.nl/isric/fulltext/isricu_i7966_001.pdf.

Lipper L, McCarthy N, Zilberman D, Asfaw S, Branca G (eds.) 2018. Climate smart agriculture. Building resilience to climate change. *Natural Resource Management and Policy* 52. Food and Agriculture Organization of the United Nations. Springer. doi:10.1007/978-3-319-61194-5_25.

Lobell, DB, Banziger M, Magorokosho C, Vivek B. 2011. Nonlinear heat effects on African maize as evidenced by historical yield trials. *Nature Climate Change* 1:42–45.

Mbata KJ, Chidumayo EN. 2003. Traditional value of caterpillars (Insecta: Lepidoptera) among the Bisa people of Zambia. *Insect Science Applications* 23(4):341–354.

Mead R, Willey RW. 1980. The concept of a land equivalent ratio and advantages in yields from intercropping. *Experimental Agriculture* 16:217–228.

Milder JC, Majanen T, Scherr SJ. 2011. Performance and potential of conservation agriculture for climate change adaptation and mitigation in Sub-Saharan Africa. Ecoagriculture Discussion Paper no. 6. Washington, DC: EcoAgriculture Partners. 91p. https://vtechworks.lib.vt.edu/handle/10919/69124.

Momoh S, Oluwalana SA, Dipeolu AO, Somefun AA. 1999. Economic analysis of rattan trade and rattan products in Lagos State. *Journal of Tropical Forest Resources* 15(1):40–51.

[NACGRAB] National Centre for Genetic Resources and Biotechnology. 2016. Varietal release catalogue 2016. Ibadan, Nigeria. www.nacgrab.gov.ng.

[NEST] Nigerian Environmental Study/Action Team, Woodley E. 2012. Learning from experience – community-based adaptation to climate change in Nigeria. NEST. Building Nigeria's Response to Climate Change (BNRCC) project Ibadan, Oyo State. 66p. http://csdevnet.org/wp-content/uploads/Learning-from-experiences.pdf.

[NNPC] Nigeria National Population Commission. 2013. Nigeria Demographic and Health Survey: Preliminary Report. Federal Ministry of Health, Abuja.

https://assets.prb.org/pdf15/nigeria-malnutrition-factsheet.pdf, 10 November 2018.

Nyasimi M, Amwata D, Hove L, Kinyangi J, Wamukoya G. 2014. Evidence of impact: climate-smart agriculture in Africa. *CCAFS Working Paper no. 86*. Copenhagen, Denmark. http://ccafs.cgiar.org/publications/evidence-impact-climate-smart-agriculture-africa.

Ofori F, Stern WR. 1987. Cereal-legume intercropping systems. *Advances in Agronomy* (41):41–90.

Okpara DA, Awurum AN, Okeke AI. 2005. Effect of planting schedule and density on cowpea/maize intercropping in south-eastern Nigeria. *Journal of Sustainable Tropical Agricultural Research* 11:59–67.

Ologunorisa T, Tersoo T. 2006. The changing rainfall pattern and its implication for flood frequency in Makurdi, Northern Nigeria. *Journal of Applied Science and Environmental Management* 10(3):97–102.

Omoigui lO, Kamara AY, Moukoumbi YD, Ogunkanmi lA, Timko MP. 2017. Breeding cowpea for resistance to *Striga gesnerioides* in the Nigerian dry savannas using marker-assisted selection. *Plant Breeding* 136(2):1–9. doi:10.1111/pbr.12475.

Omotoso OT. 2006. Nutritional quality, functional properties and anti-nutrient compositions of the larva of *Cirina forda* (Westwood) (Lepidoptera: Saturniidae). *Journal of Zhejiang University Science* 7(1):51–55. doi:10.1631/jzus.2006.B0051.

Ortese E, Baiyeri KP, Ugese FD. 2012. Demographic features of citrus producers and agronomic management of the crop in Benue State, Nigeria. *Production Agriculture and Technology Journal* 8(1):180–190.

Ouédraogo M, Partey ST, Zougmoré RB, Nuyor AB, Zakari S, Traore KB. 2018. Uptake of climate-smart agriculture in West Africa: what can we learn from climate-smart villages of Ghana, Mali and Niger? CCAFS Info Note. Bamako, Mali: CGIAR Research Program on Climate Change, Agriculture and Food Security (CCAFS). https://hdl.handle.net/10568/93351.

Parr JF, Papendick RI, Hornick SB, Meyer RE. 1990. The use of cover crops, mulches and tillage for soil water conservation and weed control. *In*: Organic matter management and tillage in Humid and Sub-humid Africa. *IBSRAM Proceedings No. 10*. IBSRAM Bangkok. P 246–261.

Partey ST, Avornyo F, Ouédraogo M, Zougmoré R. 2018. Candidate fodder species for goat production in Northern Ghana. Info note on livestock production in Ghana. Info Note. CGIAR Research Program on Climate Change, Agriculture and Food Security (CCAFS), Ghana. http://oar.icrisat.org/10697/1/Candidate%20fodder%20species%20for%20goat%20production.pdf.

Rosenstock TS, Lamanna C, Chesterman S, Bell P, Arslan A, Richards M, Rioux J, Akinleye AO, Champalle C, Cheng Z, Corner-Dolloff C, Dohn J, English W, Eyrich AS, Girvetz EH, Kerr A, Lizarazo M, Madalinska A, McFatridge S, Morris KS, Namoi N, Poultouchidou N, Ravina da Silva M, Rayess S, Ström H, Tully KL, Zhou W. 2016. The scientific basis of climate-smart agriculture: a systematic review protocol. *CCAFS Working Paper no. 138*. Copenhagen, Denmark: CGIAR Research Program on Climate Change, Agriculture and Food Security (CCAFS). https://hdl.handle.net/10568/70967.

Rosenstock TS, Nowak A, Girvetz E (eds). 2019. The Climate-Smart Agriculture Papers. Investigating the business of a productive, resilient and low emission future. Springer Open. 321p. https://link.springer.com/book/10.1007%2F978-3-319-92798-5.

Scherr S, Seth S, Friedman R. 2012*a*. From climate-smart agriculture to climate-smart landscapes. *Agriculture and Food Security* 1:12. doi:10.1186/2048-7010-1-12.

Scherr S, Shames S, Friedman R. 2012*b*. From climate smart agriculture to climate smart landscapes. *Agriculture and Food Security* (1):1–15. https://agricultureandfoodsecurity.biomedcentral.com/track/pdf/10.1186/2048-7010-1-12.

[SDSN] Sustainable Development Solutions Network. 2015. Indicators and a monitoring framework. Launching a data revolution for the Sustainable Development Goals. 233p. http://unsdsn.org/wp-content/uploads/2015/05/150612-FINAL-SDSN-Indicator-Report1.pdf.

Seran TH, Brintha I. 2010. Review on maize based intercropping. *Journal of Agronomy* 9:135–145.

Shomkegh SA, Dagba BI, Tembe ET. 2012. Tackling the challenges of forest dwelling communities towards sustainable forest management in Nigeria: a review. *In*: Onyekwelu JC, Agbeja BA, Adekunle VAJ, Lameed GA, Adesoye PO, Omole AO (eds). De-reservation, encroachment and deforestation: implications for the future of Nigerian forest estate and carbon emission reduction. proceedings of the 3rd Biennial conference of the Forest and Forest Products Society. University of Ibadan, Ibadan. P 557–581.

Shomkegh SA, Mbakwe R, Udeagha AU. 2016. Uses and relative abundance of non-timber forest products in farmlands of selected Tiv communities of Benue State, Nigeria. *Journal of Agriculture and Ecology International* 8(2):1–12. doi:10.9734/JAERI/2016/26117.

[SRP] Seasonal Rainfall Prediction. 2016. Seasonal rainfall prediction of the Nigerian Meteorological Agency (NiMet), Abuja, Nigeria. www.nimet.gov.ng/publication/2016-seasonal-rainfall-prediction-srp.

[SRP] Seasonal Rainfall Prediction. 2017. Seasonal rainfall prediction of the Nigerian Meteorological Agency (NiMet), Abuja, Nigeria. www.nimet.gov.ng/publication/2017-seasonal-rainfall-prediction-srp.

Ugese FD, Aiyeri PKB, Bah BNM. 2008. Nutritional composition of shea (*Vitellaria paradoxa*) fruit pulp across its major distribution zones in Nigeria. *Fruits* 63:163–170.

Ullah A, Bhatti MA, Gurmani ZA, Imran M. 2007. Studies on planting patterns of maize (*Zea mays* L.) facilitating legumes intercropping. *Journal of Agricultural Research* 45:113–118.

Undie UL, Uwah DF, Umoetuk SBA, Attoe EE, Effa EB. 2013. Response of late season maize/soybean intercropping to nitrogen in the humid environment of South Southern Nigeria. *Journal of Biology, Agriculture and Healthcare* 3(10):84–91.

[UNICEF] United Nations Children Fund. 2013. Improving child nutrition: the achievable imperative for global progress. New York, NY 10017 USA. 132p. www.unicef.org/gambia/Improving_Child_Nutrition_-_the_achievable_imperative_for_global_progress.pdf.

van Gelder B, O'Keefe P. 1995. The New Forester. Practical Action Publishing. London. 90p. doi:10.3362/9781780444185 https://doi.org/10.3362/9781780444185.

Victoria R, Banwart SA, Black H, Ingram H, Joosten H, Milne E, Noellemeyer E. 2012. The benefits of soil carbon: managing soils for multiple economic, societal and environmental benefits. *In*: UNEP. UNEP *Year Book 2012: Emerging Issues in Our Global Environment*. United Nations Environment Programme, Nairobi. P 19–33. doi:10.13140/2.1.1436.5125. http://wedocs.unep.org/handle/20.500.11822/8019.

Willey RW. 1985. Evaluation and presentation of intercropping advantages. *Experimental Agriculture* 21:119–133.

3 Treating shea trees as crops improves women's livelihoods in Burkina Faso

Josias Sanou with Hugues R Bazié and Jules Bayala

Current status of parkland agroforestry systems

Agroforestry parklands are land-use systems where the spaces between scattered trees and shrubs are cultivated with crops and used as pastures during the dry season (Sanou *et al.* 2004; see Figure 1.1 in Simelton, Ostwald and Osiru Chapter 1). Certain tree species provide important ecosystem services, such as water regulation, climate buffering, soil fertility, food, fodder, medicine and wood, and are kept in the field when farmers convert natural woodland to farmland (Bayala *et al.* 2014; Gijsbers *et al.* 1994).

Parkland agroforestry systems are common in West African Savanna and have been used by farmers to obtain products from both trees and annual food crops, particularly for food security. The tree density and species composition are determined by the value to farmers of the products and services provided by trees. The parklands in the sub-humid zone of West Africa are mainly composed of shea (*Vitellaria paradoxa* C. F. Gaertn), néré *(Parkia biglobosa* (Jacq.) R. Br. ex G. Don) and *Faidherbia albida*, while in semi-arid areas, the dominant trees in parklands are acacias (*Acacia raddiana, Acacia Senegal*), *Adansonia digitata* and desert date (*Balanites aegyptica*) (Boffa 1999). Parklands show a great diversity, and while shea is the single most common tree species covering over 20 per cent of the Sudanian and South-Sahelian ecozones of Burkina Faso, the remaining 80 per cent consists of 40 other species (Nikiéma 2005; Figure 1.5 in Simelton, Ostwald and Osiru Chapter 1). These agroforestry parklands are multifunctional landscapes in the sense that they play multiple roles in rural livelihoods and food provision – allowing for the integration with cereals (maize, sorghum and millet), roots and tubers (yam, sweet potatoes and cassava), legumes (cowpea, peanut and Bambara groundnut) and vegetables (sorrel, okra, chilli pepper, eggplant). A review of West African agroforestry parklands and woody amendments showed that overall, the presence of shrubs and trees in parklands improved soil carbon, millet, and sorghum yields (Félix *et al.* 2018). Also, the trees themselves provide diverse sources of fruits, fats, oils, leafy vegetables, nuts and condiments that supply micro-nutrients and vitamins to complement the typical cereal-based diets (Bayala *et al.* 2014). On a larger scale, shea trees buffer against desertification while the incomes from shea butter help make households more economically resilient to adverse climatic events (Hammond *et al.* 2019).

Before 1970, rural landscapes in Burkina Faso were divided into three uses: farmland, fallow land and natural woodland. Crops were grown on farmland for three to ten years depending on soil fertility, and when soil fertility was too low, land was left fallow. After five to 15 years, land left fallow would have regrown into woodland, restored the soils, and now be ready for a new cycle of crop cultivation (Boffa 1999). However, in the middle of the 1990s, researchers noticed a declining trend in tree density and poor performance of parkland production systems (Boffa *et al.* 1996). This was caused by a combination of factors. Notably, the droughts were becoming more frequent. Population growth made it harder to leave land fallow, reducing the period to two to three years and increasing the time between fallow periods, resulting in a decline in soil fertility. Moreover, as dead and ageing regenerated trees were not replaced, fewer trees remained in the landscapes (Maranz 2009; Ræbild *et al.* 2012).

Against this background, actions were initiated to restore the parklands and increase the production and productivity of ecosystem services. Projects aimed to enhance interactions between trees and associated crops, tree regeneration and system management (Bayala *et al.* 2014; Bazié *et al.* 2012; Gijsbers *et al.* 1994; Ouédraogo 1994; Sanou *et al.* 2012). Assessments of these activities revealed that shade was the main cause of the yield decrease of the associated crops, particularly for cereals, such as millet and sorghum (Bayala *et al.* 2013). However, Jonsson *et al.* (1999) reported a positive effect on millet yields as trees reduced temperatures by providing shade, which improved soil moisture. Such divergence might be linked to differences in the rainfall patterns, as trees can buffer sparsely distributed rains during years with dry spells (Bayala *et al.* 2008, 2014).

To resolve the problem of tree shade, branches of old trees were totally pruned for rejuvenation, while partial pruning was recommended for removing parasites such as *Tapinanthus spp* (Boussim *et al.* 1993*b*). Bayala (2002) tested this approach on shea and néré with good results. Cereal crop yields increased by 400 to 808 per cent and soil fertility improved by using the pruned leaves for mulching (Bayala *et al.* 2002). One drawback was observed. The total pruning of trees for rejuvenation reduced tree fruit production for at least five years until the trees recovered to their original production level (Bayala *et al.* 2008). However, the lost fruit yields were compensated for by increased crop production under pruned trees. In contrast, if farmers are not allowed to prune trees, one alternative would be to plant shade-tolerant crops beneath tree canopies and sun-loving crops outside the shaded area (Nur Osman *et al.* 2011; Pouliot *et al.* 2012; Sanou *et al.* 2012).

Shea agroforestry parkland in the Sudanese zone of Burkina Faso

The main land-use system in the Sudanese zone of Burkina Faso is shea agroforestry parkland. The system is distributed across almost the whole country, albeit with a higher density of more than 45 trees per hectare in the Sudanese zone. The shea parklands system mixes well with other woody species such as néré, baobab (*Adansonia digitate*), tamarind (*Tamarindus indica*), desert date and *Lannea microcarpa*, which can be intercropped with subsistence and cash crops (Figure 3.1) and used for pasture during the dry season.

The land owner and the land user are not necessarily the same person. This is particularly the case for most women, where, traditionally, land is owned by men while women can access the land to harvest shea trees. In this case, women and men who are not land owners are forbidden to plant any species of trees and harvest certain rare tree species, néré being one of them.

Figure 3.1 Shea parkland with millet and maize in Nobéré, Burkina Faso, West Africa.

Photo credit: Sanou 2017.

Food crops production on shea agroforestry parklands

Shea agroforestry parkland fields are normally up to 20 hectares of diverse food crops. Farmers often integrate cereal crops with a legume, such as peanut or cowpea. Cash crops, such as cotton and sesame, can also be grown in these systems. Vegetables, such as sorrel, okra and eggplants, are often found as the sole crop on smaller plots around the homestead areas, under large canopy trees or in rows, intercropped with cereals.

Most farmers keep goats and sheep, sometimes cattle. Crop residues are collected for animal feed, and during the dry season farmland and fallow land are used as pasture. Manure is collected and used to fertilize parkland soils. Farmers without livestock sometimes let herders use their parkland for pasture during the dry season, hence benefitting from the manure.

Shea tree products and functions

Shea trees provide firewood and fruits and host shea caterpillars (*Cirina butyrospermii*) and beekeeping (Vodoude et al. 2009). The pulp of the shea fruit is a common source of vitamins A and B (Boffa et al. 1996; Hall et al. 1996; Ugese et al. 2008). The butter from shea nuts is used for cooking (Figure 3.2) and makes up 88 per cent of the fat and oils consumed by rural households, and 25 per cent of that consumed by urban households in Burkina Faso. In 2011, Burkina Faso's total consumption of shea butter amounted to 11,826 tonnes, equal to 1.5 kilogrammes per capita (INSD 2011). Shea butter is also sold for industrial purposes, exported and used to replace cocoa butter in chocolate and in cosmetics (Fold and Reenberg 1999). According to FAO (2011), shea products were the country's fourth-largest export commodity after gold, cotton and livestock. In 2015, shea exports were valued at US$44.5 million (APEX 2016).

Traditionally, shea caterpillars were mainly consumed in western Burkina Faso but are now widely consumed across the country, and they are the second-most eaten and sold edible insect in Burkina Faso, Mali and Togo (Tchibozo et al. 2016). Shea caterpillars are an important source of protein, especially for children. Due to the nutritional value, demand has increased. With growing markets, the caterpillar has become a source of income for women, who are the main collectors and traders. Sermé (2011) estimated on average the net annual revenue earned by collectors to be above US$300, which is equivalent to the annual fees for two pupils in secondary school.

Finally, shea trees are important for honey production. The nectar-producing flowers attract bees, which has made the tree a preferred location for keeping traditional bee hives (Sallé et al. 1991).

Similar to other tree-based systems, shea parklands regulate the microclimate (Bayala et al. 2014). For example, shea can reduce the

Figure 3.2 Shea butter processed and traded by a woman in Nobéré, Burkina Faso.
Photo credit: Sanou 2017.

ambient mean temperature by up to one degree Celsius compared to open spaces (Jonsson *et al.* 1999) and reduce the wind speed, which increases the soil and air humidity by up to 5 per cent (Bayala *et al.* 2014). The accumulated biomass from tree litter and root decomposition helps improve soil properties (Bayala *et al.* 2014). For example, the increase in soil moisture is explained by the hydraulic lift mechanism, whereby water is driven from deeper soil layers up to the topsoil, as the trees help improve soil hydraulic properties and contribute to groundwater recharge in the parklands (Bargués Tobella *et al.* 2014; Ilstedt *et al.* 2016). Furthermore, through photosynthesis, shea trees sequester carbon in the trunk and roots at 8.9 tonnes carbon per hectare, on average (Shu-aib Jakpa 2016).

Importance of shea for rural women's livelihoods

Shea is sometimes called 'women's gold' because it is the most important income source for rural women, contributing up to 12 per cent of poor-household incomes (Elias and Carney 2007; Hammond *et al.* 2019). In Burkina Faso, women are mostly in charge of both collecting and processing the nuts (Pouliot 2012). The emergence of international markets since the 1980s has offered a unique opportunity for rural women, who otherwise have few income-generating activities (Compaoré 2000). However, this opportunity depends on two critical factors. First, land tenure – maintaining traditional collector access to shea nuts and trees in parklands. Second, production costs – reducing the labour involved in processing butter from nuts (Ouedraogo 2012). The traditional process requires energy. Yokabdjim (2006) estimated that it takes nearly nine hours for one woman to process ten kilogrammes of shea nuts into about two kilogrammes of butter.

Stakeholders in shea production on agroforestry parklands

Due to the income opportunity, several governmental projects on shea collection, processing and trading were implemented in Burkina Faso with a focus on empowering women (Badini *et al.* 2011), including a national strategy for sustainable development of the shea industry (2015–2019). The challenge of this strategy is to respond to market requirements, while considering the socioeconomic and environmental impacts of the shea industry. The extension services of both the ministries of environment and

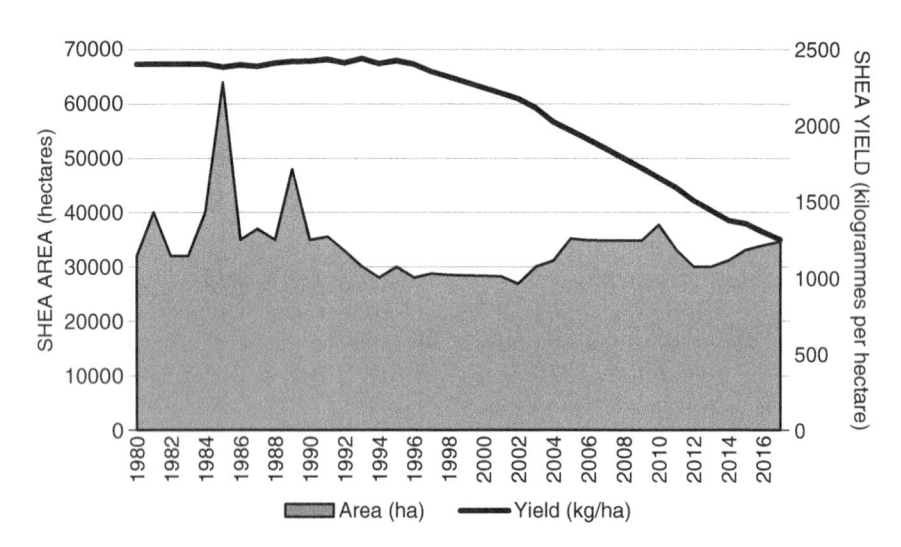

Figure 3.3 Shea nut productivity and harvested area in Burkina Faso from 1980 to 2017.

Source: FAOSTAT 2018.

agriculture are assisting farmers in the management of shea parkland agroforestry systems. Furthermore, non-governmental organizations and the national agency for non-wood forest products promotion are concerned with shea conservation and training women on nuts processing; and in 2011, the agency recorded 1069 professional organizations collecting, processing and trading shea nuts in Burkina Faso (APFNL 2010).

Management practices for shea trees

From 1970, when the cycle of longer fallow periods was broken, soil fertility declined, the average age of the trees increased and the trees were increasingly attacked by the parasite *Tapinanthus spp*. The permanent cropping of parklands did not allow for natural soil regeneration. As shea trees are not traditionally planted, skipping fallow periods resulted in tree-ageing in parklands. The *Tapinanthus spp* infestation rate was higher in parkland trees compared to forests, because of birds spreading the mistletoe to isolated trees (Boussim *et al.* 1993a). The soil fertility decline resulted in lower yields from annual crops in general, while ageing pest-ridden trees reduced their yield.

Starting in 2000, parkland management was proposed in order to maintain good productivity through several projects under the Institut de l'Environnement et de Recherches Agricoles. For example, the SAFRUIT project was implemented in Burkina Faso, Mali and Niger and supported germplasm and local knowledge about fruit trees. BIODEV focussed on development benefits from biological and carbon stock improvements through agroforestry, forest management and tree planting. The McKnight-AEI project focussed on agroecological intensification of sorghum and pearl millet through agroforestry in the Sahel. INNOVKAR was a research project on innovative shea tree techniques. Management included tree planting, pruning trees for rejuvenation or health, and assisted natural regeneration (Bayala *et al.* 2008).

Shea rejuvenation is a slow process, as a tree will start fruiting after 15–20 years. Grafting is promoted to shorten the vegetative phase (Sanou *et al.* 2004). Branch debarking has been successfully tested to improve flowering and fruiting (Lamien *et al.* 2006). Both total and partial pruning increased shea fruit production. For example, old shea trees produce a maximum of five kilogrammes per year compared to on average 20 kilogrammes six years after pruning, which is comparable to a young tree (Bayala *et al.* 2008). Total pruning, in which all branches are removed on trees that no longer bear fruit, is a practice used to rejuvenate old trees. Total pruning (Figure 3.4) increased millet yield underneath the tree by 300 per cent (Bayala *et al.* 2002). In another study, pruning the crowns increased sorghum yields by 520 per cent and straw dry matter biomass by 348 per cent, as more sunlight penetrated and soil fertility increased under the pruned trees (Bazié *et al.* 2012). Partial pruning is done by removing

Figure 3.4 Pruned and unpruned shea trees on a parkland in Nobéré, Burkina Faso, West Africa.
Photo credit: Sanou 2017.

branches infested by African mistletoe to improve tree health. Fruit production in totally pruned trees recovered by 80 per cent within five years after pruning and fully (100 per cent) after six years.

The assisted natural regeneration, also known as farmer-managed natural regeneration, is an agroforestry approach that consists of keeping and maintaining young spontaneous plants of woody species at desired densities in the cultivated plot (Reij and Garrity 2016; Sacande and Berrahmouni 2016). The practice helps regenerate shea trees and increase tree density and can be combined with grafting to improve the nut quality and shorten the time to the first harvest from 15 to five years (Sanou *et al.* 2004).

Factors for success in studied cases

Pruning, assisted natural regeneration and grafting techniques are affordable practices, but require skilled labour. Generally, farmers already have the equipment, such as axes and machetes, and need training on how and when to prune branches, for instance to avoid pruning during the rainy season as rainwater can cause rotting of wounded branches or kill the tree.

Keeping and managing shea trees on parklands resulted in an increase in fruit production and an increase in incomes for rural women. A survey of 150 households, 64 of which were female-headed, was conducted in Nobéré in 2016. Half of the respondents had taken part in the training on tree management and had been practising it since 2010. The study was conducted as a field visit to assess trees and soil information followed by a discussion with households for information on tree management practices, tree and crop production and household revenues. The survey results showed that average revenue earned from shea nut sales was US$108, which was 44 per cent higher in the group that practised pruning and assisted natural regeneration (US$6 per tree compared to US$4 per tree). The revenue from wood production was also higher with the tree management practices, on average US$12 compared to US$3 per hectare. Studies in other West African countries also point to the particular benefits of shea to women's livelihoods (Elias and Carney 2007; Faye *et al.* 2010; Hammond *et al.* 2019).

Limitations

The rural population of Burkina Faso tripled between 1960 and 2014. Meanwhile, especially since 2000, the cropland increased from about 300,000 to 580,000 hectares, which resulted in the average land per capita remaining fairly stable at 0.40 to 0.55 hectares (FAOSTAT 2018). However, farmers tend to reclaim all their land for crop production, such as fallow land, degraded land and space under large trees, and eliminate non-productive trees that interfere with the yield of cereal crops, for example by shade. This is one of the main causes of the reduction in tree density on agroforestry parklands (Sanou *et al.* 2012).

Second, reduced tree density depended on the parasitic African mistletoe (*Tapinanthus spp*) and high number of old trees with low fruit production (Lamien *et al.* 2004; Traore *et al.* 2007). According to Boussim *et al.* (1993*a*), 95 per cent of shea trees in Burkina Faso were affected by the three African mistletoe species of the Loranthaceae family: *Agelanthus dodoneifolius* (DC.) Polh. & Wiens, *Tapinanthus globiferus* (A. Rich) Van Tieghem and *Tapinanthus ophiodes* (Sprague) Dansers. These mistletoes are spread by birds who feed on their seeds and regurgitate them onto new trees where they germinate, project through the bark and, as they grow, cause discontinued growth, withering of tree parts and finally tree mortality.

The shea nut yield in Burkina Faso indicates a high variability year to year and a decrease within the last decade, ranging between five and eight tonnes per hectare.

Limits to further expansion of Shea parklands management

The further expansion of shea parklands management practices depends on three main aspects. First, in most West African countries, agroforestry parkland trees fall under forest legislation, which prohibits unauthorized tree management even when farmers are the land owners. In addition, in Burkina Faso, the shea tree is listed as a protected species (MECV 2014). As such, it may not be cut, torn down, mutilated or incinerated without permission from forest authorities. Due to these regulations, farmers tend to reduce tree density when trees are unproductive simply by not replacing dead or dying trees. Moreover, because of the insecure tenure status, land users who are not land owners do not regenerate trees.

Second, assisted natural regeneration presents difficulties in protecting shea seedlings during the dry season. Irrigation requires transporting water long distances, which most farmers cannot afford. As parklands are cultivated in the wet season and used as pastureland in the dry season, some farmers are discouraged from keeping tree seedlings as they hinder ploughing with animals and young plants risk being grazed by roaming animals.

Third, biophysical limitations include climate variability, particularly droughts and bush fires, which reduce regeneration and cause trees mortality. Additionally, land pressure caused by population growth and urban development reduces the area of parklands.

Policy aspects

The success of shea parklands management depends on policies that:

1. ensure long-term land ownership, especially for women, and secure user rights to trees and their products;
2. include management, such as pruning for tree health or rejuvenation, and encourage new regrowth through assisted natural regeneration, planting and grafting;
3. enhance women's skills in processing nuts to make value-added shea products to improve incomes and provide access to the international market for shea butter.

Lessons learned from the case

Parkland agroforestry systems are essential for the livelihoods of rural people in semi-arid West Africa. Good management of trees on parklands enhances food security, conserves environment for sustainable production and increases farmer revenues especially for rural women. It is necessary to revise legislation on agroforestry parklands to secure long-term tree management practices to sustain productivity of the systems, such as those requiring authorization from the forest department for tree management.

Furthermore, research is needed to increase understanding of the tree-crop interactions and trade-offs in shea parklands systems, and the ways in which these are modified by specific social and ecological contexts.

References

[APEX] Agence pour la promotion des Exportations du Burkina. 2016. Offre exportable du Burkina Faso: cas du karité No 1. Ougadougou. 2p. www.apexb.bf/wp-content/uploads/Telecharger-fiche-karite-burkina.pdf.

[APFNL] Agence des Produits Forestiers Non Ligneux. 2010. Stratégie nationale de valorisation et de promotion des produits forestiers non ligneux. Agence des Produits Forestiers Non Ligneux. Ministere de l'Environnement Burkina Faso et du Cadre de Vie, Ouagadougou, Burkina Faso. 74p. www.fao.org/forestry/32339-07692e83cf2f5f62453d9b2d074c4df6c.pdf.

Badini Z, Kabore M, Mheen-Sluijer J, van der Vellema S. 2011. Historique de la filière karité au Burkina Faso et des services offerts par les partenaires techniques et financiers aux acteurs. VC4PD Research Paper No. 11. Wageningen University and Research Centre. Wageningen. http://edepot.wur.nl/187139.

Bargués Tobella A, Reese H, Almaw A, Bayala J, Malmer A, Laudon H, Ilstedt U. 2014. The effect of trees on preferential flow and soil infiltrability in an agroforestry parkland in semiarid Burkina Faso. *Water Resources Research* 50:3342–3354.

Bayala J. 2002. Tree crown pruning as a management tool to enhance the productivity of parklands in West Africa. PhD thesis. University of Wales, Bangor.

Bayala J, Teklehaimanot Z, Ouédraogo SJ. 2002. Millet production under pruned tree crowns in a parkland system in Burkina Faso. *Agroforestry Systems* 54:203–214.

Bayala J, Bazié HR, Sanou J. 2013. Competition and facilitation-related factors' impacts on crop performance in an agro-forestry parkland system in Burkina Faso. *African Journal of Agricultural Research* 8(43):5303–5310. doi:10.5897/ajar11.1843.

Bayala J, Sanou J, Teklehaimanot Z, Kalinganire A, Ouédraogo SJ. 2014. Parklands for buffering climate risk and sustaining agricultural production in the Sahel of West Africa. *Current Opinion in Environmental Sustainability* 6:28–34.

Bayala J, Ouédraogo SJ, Teklehaimanot Z. 2008. Rejuvenating indigenous trees in agroforestry parkland systems for better fruit production using crown pruning. *Agroforestry Systems* 72:187–194. doi:10.1007/s10457-007-9099-9.

Bazié HR, Bayala J, Zombré G, Sanou J, Ilstedt U. 2012. Separating competition-related factors limiting crop performance in an agroforestry parkland system in Burkina Faso. *Agroforestry Systems* 84:377–388.

Boffa JM. 1999. Agroforestry parklands in Sub-Saharan Africa. Conservation Guide 34. Food and Agriculture Organization of the United Nations, Rome. www.fao.org/3/x3940e/X3940E00.htm.

Boffa JM, Yaméogo G, Nikiéma P, Knudson DM. 1996. Shea nut (*Vitellaria paradoxa*) production and collection in agroforestry parklands of Burkina Faso. *In*: Leakey RRB, Temu AB, Melnyk M, Vantomme P (eds). *Domestication and Commercialization of Non-Timber Forest Products in Agroforestry Systems*, Non-wood Forest Products 9:110–122. Food and Agriculture Organization of the United Nations. Rome. P 110–122. www.fao.org/3/a-w3735e.pdf.

Boussim IJ, Sallé G, Guinko S. 1993a. Tapinanthus, parasite du Karité au Burkina Faso. Identification et Distribution 1ère partie. *Bois et Forêts des Tropiques* 238:45–52.

Boussim IJ, Sallé G, Guinko S. 1993b. *Tapinanthus* parasite du karité au Burkina Faso. 2éme partie: Phénologie, biologie et dégâts. *Bois et Forêts des Tropiques* 238:53–65.

Compaoré N. 2000. Femmes, développement et transfert de technologies. Le cas des presses à karité au Burkina Faso. Thèse de doctorat, Université de Montréal, Canada.

Elias M, Carney J. 2007. African Shea butter: a feminized subsidy from nature. *Africa* 77(11):37–62. doi:10.3366/afr.2007.77.1.37.

[FAO] Food and Agriculture Organization of the United Nations. 2011. CountrySTAT Burkina Faso. Structure des exportations par Désignation et Valeur. Rome. http://countrystat.org/bfa/cont/pxweb query/ma/233itr015/fr.

FAOSTAT 2018. Food and Agriculture Organization of the United Nations. Rome. www.fao.org/faostat/en/#data/QC.

Faye MD, Weber JC, Mounkoro B, Dakouo JM. 2010. Contribution of parkland trees to farmers' livelihoods: a case study from Mali. *Development in Practice* 20(3):428–434. doi:10.1080/09614521003710013.

Félix GF, Scholberg JMS, Clermont-Dauphin C, Laurent Cournac L, Tittonell P. 2018. Enhancing agroecosystem productivity with woody perennials in semi-arid West Africa. A meta-analysis. *Agronomy for Sustainable Development* 38:57. https://doi.org/10.1007/s13593-018-0533-3.

Fold N, Reenberg A. 1999. In the shadow of the 'Chocolate War': local marketing of Shea nut products around Tenkodogo, Burkina Faso. *Geografisk Tidsskrift* 2:113–124.

Gijsbers HJM, Kessler JJ, Knevel MK. 1994. Dynamics and natural regeneration of woody species in farmed parkland in the Sahel region (Province of Passoré) Burkina Faso. *Forestry Ecology Management* 64:1–12.

Hall JB, Aebischer DP, Tomlinson HF, Osei-Amaning E, Hindle JR. 1996. Vitellaria paradoxa: a monograph. Publication No. 8. School of Agricultural and Forest Sciences, University of Wales, Bangor. 105p.

Hammond J, van Wijk M, Pagella T, Carpena P, Skirrow T, Dauncey V. 2019. Shea Butter: a pro-poor, pro-female route to increase income. *In*: Rosenstock T, Nowak A, Girvetz E (eds). The Climate smart agriculture papers. Investigating the business of a productive, resilient and low emission future. *Springer Open*. P 215–226. https://link.springer.com/chapter/10.1007/978-3-319-92798-5_18.

Ilstedt U, Bargués Tobella A, Bazié HR, Bayala J, Verbeeten E, Nyberg G, Sanou J, Benegas L, Murdiyarso D, Laudon H, Sheil D, Malmer A. 2016. Intermediate tree cover can maximize groundwater recharge in the seasonally dry tropics. *Scientific Reports* 6:21930; doi:10.1038/srep21930.

[INSD] Institut Nationale de la Statistique et de la Démographie. 2011. Analyse des Résultats provisoires de l'Enquête Intégrale Dépenses des Ménages. INSD, Ouagadougou, Burkina Faso.

Jonsson K, Ong CK, Odongos JC. 1999. Influence of scattered néré and karité' on microclimate, soil fertility and millet yield in Burkina Faso. *Experimental Agriculture* 35(24):39–53.

Lamien N, Ouédraogo SJ, Diallo OB, Guinko S. 2004. Productivité fruitière du karité (*Vitellaria paradoxa* C. F. Gaertn, Sapotaceae) dans les parcs agroforestiers traditionnels au Burkina Faso. *Fruits* 59:423–429.

Lamien N, Tigabu M, Oden PC, Guinko S. 2006. Effets de l'incision annulaire sur la reproduction du karité (*Vitellaria paradoxa* C.F. Gaertn.) à Bondoukuy, Burkina Faso. *Fruits* 61:303–312.

Maranz S. 2009. Tree mortality in the African Sahel indicates an anthropogenic ecosystem displaced by climate change. *Journal of Biogeography* 36:1181–1193.

[MECV] Ministere de l'Environnement et du Cadre de Vie. 2014. Arrêté n°2004–019/MECV portant détermination de la liste des espèces forestières bénéficiant de mesures de protection particulière. MECV. Ouagadougou. Burkina Faso.

Nikiéma A. 2005. Agroforestry parkland species diversity: uses and management in semi-arid West Africa (Burkina Faso). PhD thesis Wageningen University, Wageningen.

Nur Osman A, Ræbild A, Christiansen JL, Bayala J. 2011. Performance of cowpea (*Vigna unguiculata*) and pearl millet (*Pennisetum glaucum*) intercropped under *Parkia biglobosa* in an agroforestry system in Burkina Faso. *African Journal Agricultural Research* 6(4):882–891.

Ouedraogo B. 2012. Etude sur la contribution des PFNL à l'économie nationale: cas de la filière fruit de l'arbre à karité. Rapport final. MEDD Burkina Faso. doi.org/10.13140/rg.2.1.3497.0969.

Ouédraogo SJ. 1994. Dynamique et fonctionnement des parcs agroforestiers traditionnels du Plateau Central burkinabé. Influence des facteurs biophysiques et anthropiques sur la composante arborée. Thèse de l'Université. Paris 6.

Pouliot M. 2012. Contribution of 'Women's Gold' to West African Livelihoods: The Case of Shea (*Vitellaria paradoxa*) in Burkina Faso. *Economic Botany* 66(3):237–248.

Pouliot M, Bayala J, Ræbild A. 2012. Testing the shade tolerance of selected crops under *Parkia biglobosa* (Jacq.) Benth. in an agroforestry parkland in Burkina Faso, West Africa. *Agroforestry Systems* 85:477–488.

Ræbild A, Hansen UB, Kambou S. 2012. Regeneration of *Vitellaria paradoxa* and *Parkia biglobosa* in a parkland in southern Burkina Faso. *Agroforestry Systems* 85:443–453.

Reij C, Garrity D. 2016. Scaling up farmer-managed natural regeneration in Africa to restore degraded landscapes. *Biotropica* 48(6):834–843.

Sacande M, Berrahmouni N. 2016. Community participation and ecological criteria for selecting species and restoring natural capital with native species in the Sahel. *Restoration Ecology* 24(4):479–488.

Sallé G, Boussim J, Raynal-Roques A, Brunck F. 1991. Le karité: une richesse potentielle. Perspectives de recherche pour améliorer sa production. *Bois et Forêts des Tropiques* 228:11–23.

Sanou H, Kambou S, Teklehaimanot Z, Dembélé M, Yossi H, Sina S, Lompo D, Bouvet JM. 2004. Vegetative propagation of Vitellaria paradoxa by grafting. *Agroforestry systems* 60:93–99.

Sanou J, Bayala J, Teklehaimanot Z, Bazié P. 2012. Effect of shading by baobab (*Adansonia digitata*) and néré (*Parkia biglobosa*) on yields of millet (*Pennisetum glaucum*) and taro (Colocasia esculenta) in parkland systems in Burkina Faso, West Africa. *Agroforestry Systems* 85:431–441.

Sermé B. 2011. Le circuit de commercialisation et importance économique des chenilles de karité (*Cirina butyrospermii*) dans la province du Houet. Mémoire de fin de cycle IDR, Université Polytechnique de Bobo-Dioulasso, Burkina Faso. 48p.

Shu-aib Jakpa S. 2016. Global to local sustainability: international climate change agreements and shea production in Ghana. Ph.D. Thesis, University of Twente, Netherlands. 247p.

Tchibozo S, Malaisse F, Mergen P. 2016. Insectes consommés par l'Homme en Afrique occidentale francophone. *Geo-Eco-Trop* 40–2:105–114.

Traore O, Some AN, Traore K, Nikiéma P, Some K. 2007. State and trends of woody vegetation cover in the cotton-based farming system zone of Western Burkina Faso. *International Journal of Biological and Chemical Sciences* 1(1):43–53.

Ugese FD, Aiyeri PKB, Bah BNM. 2008. Nutritional composition of shea (*Vitellaria paradoxa*) fruit pulp across its major distribution zones in Nigeria. *Fruits* 63:163–170.

Vodoude FG, Coulibaly O, Greene C, Sinsin. B. 2009. Estimating the local value of non-timber forest products to Pendjari Biosphere Reserve dwellers in Benin. *Economic Botany* 63(4):397–412.

Yokabdjim P. 2006. Etude de plateforme-multifonctionnelle: services énergétiques et processus de développement humain durable; le cas de la filière karité dans le centre-Ouest du Burkina Faso. Mémoire de fin d'étude d'ingénieur de l'équipement rural. Groupe des écoles EIER-ETSHER, Burkina Faso. 77p. http://documentation.2ie-edu.org/cdi2ie/opac_css/doc_num.php?explnum_id=942.

4 Economic benefits from cassava in peri-urban multiple-cropping systems in Nigeria

Anthony Ojonimi Onoja

Current state of cassava

Introduction

Two converging trends form the backdrop for this chapter. First, in the 2000s, the global urban population exceeded the rural for the first time (Satterthwaite *et al.* 2010). Second, Africa's most populous country, Nigeria, is expected to double its population to 410 million in 20 years. This makes Nigeria's food security relevant in the context of both global population growth (UNDESA 2015) and the millions of poor urban citizens who will be unable to grow their own food (Satterthwaite *et al.* 2010).

Nigeria's recent agrarian history matches that of a so-called 'resource-cursed' country, with investments in agriculture and rural development being neglected while food-import dependency is built up driven by the oil and gas industry. Nigeria is one of Africa's two largest economies and income inequality is high, with the Gini coefficient peaking at 0.52 in 1996 and most recently estimated at 0.43 in 2009 (Bakare 2012; WB 2018), and all the higher in urban areas (a Gini coefficient of 1, or 100 per cent, expresses maximal inequality). Nigeria is a net food importer, and food is expensive for many. The food-insecure population is projected to more than double from 17 million in 2012 to 43 million in 2022 (FAO 2018a; MBNP 2016). By 2014, the jihadist militant organization Boko Haram had displaced 1.6 million Nigerians, many of whom are now unable to grow their food, increasing the risk of urban hunger. The same period has seen a rural–urban migration trend (Métivier 2015).

The role of cassava in Sub-Saharan Africa

Cassava is one of the main staple foods in Sub-Saharan Africa, one of the two most important staples in Nigeria, providing at least one-third of the calorie intake and a much larger share among the poor (De Souza *et al.* 2016). Estimates suggest that urban Nigerians eat 200 grammes of cassava per day (FAO 2013).

Nigeria has become the world's largest producer of cassava, at over 60 million tonnes, followed by Thailand and the Democratic Republic of Congo, each producing about 30 million tonnes (FAOSTAT 2018). As with maize (Adewopo Chapter 7), cassava production levels in Nigeria were achieved through expanding harvested areas, investing in processing infrastructure, and choosing high-yielding varieties (FAO 2013; IFAD and FAO 2005). Despite this, average cassava yields in Nigeria, as in many African countries, have remained at the same levels for the past decades, while the area has increased (Figure 4.1). Meanwhile, the yield has steadily increased in Thailand, while the area has remained constant. This could depend on Thailand investing in developing high-yield varieties for various purposes, while in Nigeria cassava is largely grown by smallholder farmers with more mixed farming systems (FAO 2013), resulting in lower yields.

Relatively tolerant to drought, nutrient-poor soils, and pests, cassava is suitable across most of the semi-tropics and tropics. About two-thirds of Nigeria's cassava production is from the southern states, where yields are the highest (FAO 2018a,b). However, concerns about the production levels that can be achieved are heightened by the exposure of degraded, rainfed farmlands to climate change. Moreover, Ropo and Ibraheem (2017) found that cassava yields in Port Harcourt were particularly sensitive to increasing minimum temperatures compared to, for example, increasing variability in rainfall and maximum temperatures.

Cassava itself provides multiple products: tuberous roots and nutritious leaves that are used for human consumption and animal feed and stems

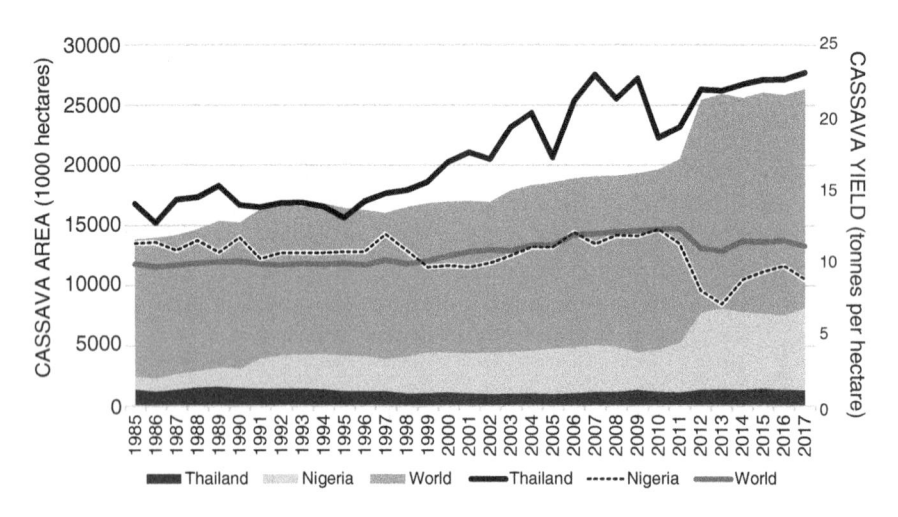

Figure 4.1 Area planted with cassava for Nigeria and Thailand as share of the global area (stacked), and yields (lines) 1985 to 2017.

Source: FAOSTAT 2018.

that can be used as fence, green mulch, or fuel for cooking. In general, multiple-cropping systems have quite consistently delivered more benefits than monocultures in Africa (Mander et al. 2007; Mbow et al. 2014). In terms of income generation in the rural parts of the central Niger Delta, cassava-based multiple-cropping systems were superior to monoculture (Ajayi 2014; Allison-Oguru et al. 2006). Specifically, while the profit from monoculture cassava in the Niger Delta was about US$860 (NGN23,700) per hectare, the cassava multiple-cropping system (cassava, maize and vegetable) recorded a profit of US$1,330 (NGN36,650) per hectare, resulting in an economic Land Equivalent Ratio (see Figure 1.3 in Simelton, Ostwald and Osiru Chapter 1) of 1.59 (Bamiro et al. 2012). In response to the food crisis, urban dwellers in Port Harcourt have diversified and developed home gardens (Onoja and Ajie 2015). However, few studies have focussed on peri-urban cassava farmers or those in the urban core.

Cassava-based multiple-cropping system in Nigeria

This case study targets peri-urban cassava-based farming systems in Port Harcourt, Nigeria's third largest commercial centre after Lagos and Kano (Figure 1.5 in Simelton, Ostwald and Osiru Chapter 1). These systems are important to household food security. The study identifies major drivers for adoption of different cassava-based multiple-cropping practices and the associated benefits to the livelihoods of peri-urban smallholders. Here, 'multiple-cropping practices' refers to practices where cassava is grown with other annual or perennial crops, including intercropping, cover crops, crop rotation and agroforestry (Table 4.1).

Table 4.1 Peri-urban cassava-based multiple-cropping systems

Cropping pattern	Description per land parcel
Intercropping, cover crops	Different crops planted randomly or in alternate rows, to minimise competition between crops and maximise soil moisture and nutrient uptake. Nitrogen-fixing plants can function as cover crops to reduce weeds and prevent erosion.
Crop rotation, relay cropping	Two or more crops grown in sequence to make use of soil nutrients.
Agroforestry	Annual crops mixed with shrubs, perennial fruit, or timber trees. The deeper-rooted trees can often draw water and nutrients that are otherwise unavailable to the crops while the trees may equally provide shade and mulch. Can be combined with cover crops.

Source: Adapted from TMP (2013).

Cassava in peri-urban multiple-cropping systems: A case study

Study site

This case study on cassava-based multiple-cropping systems was conducted in the peri-urban fringes of Port Harcourt, the capital of Rivers State in the Niger Delta in the South-South zone of Nigeria. In 2016, the state's population reached seven million (NBS 2018; RSG 2018), of which 25 per cent reside in Port Harcourt. Its 190,000 hectares were traditional farmland until crude oil was discovered, and Port Harcourt became the centre for the oil industry in the 1960s. Since then, the city has experienced a rapid influx of job migrants with a population increase of approximately 35 per cent between 2006 and 2016. This increase has put pressure on agricultural land (Satterthwaite *et al.* 2010), where farmers have left agriculture or have been forced into more intensive production on the remaining farmland and encroached on non-agriculture land (Nlerum and Wechie 2018). About 70 per cent of the population relies on tropical crops, such as cassava, yam, maize, potato, pineapple, vegetables, plantain and banana, alongside forestry (RSG 2018) and hunting wild animals, such as civets, marshbucks and antelopes.

Data

Primary data were collected from a field survey of 150 farm households in three peri-urban areas of Port Harcourt: Obio-Akpor, Etche and Ekwerre. The survey consisted of a structured questionnaire for quantitative assessment and focus group discussions for further clarification and for qualitative analysis of environmental benefits. In all, 75 women and 75 men were randomly selected from the Agricultural Development Project's list of registered farmers in Port Harcourt. Unconfirmed data suggest that as many as 87 per cent of the farms in Port Harcourt are headed by women. The 150 interviews were conducted in November 2017. The questionnaire also solicited some responses on farm practices prior to 2010.

The farm gains from different multiple-cropping systems can be expressed as the gross margin equivalent to the farm revenues minus variable costs (Kahan 2013). Production costs are unique to each farming operation, which may cause some disagreements about what costs to include. Total cost normally includes variable and fixed costs, where variable costs vary with output within a production period and result from the use of purchased inputs and owned assets, and fixed costs do not vary with the level of output and result from ownership of assets (Samuelson and Nordhaus 2005). Here, variable costs included costs for planting materials (stem cutting), labour (hired labour and estimated cost for family labour), fertilizer (including manure and inorganic fertilizers), pesticides, transport expenses, operation, interest on loans received, land rents, and cost of packaging. In this case, fixed costs such as the depreciated value of farm tools and equipment (cutlasses, hoes, pans, rakes and wheel barrows) were marginal and not included in the analysis. Productivity was assessed based on the cassava yield.

The net income was estimated by deducting the total costs from the 'total revenue', that is the monetary value of total harvests including home consumption (the conversion rate was Nigerian Naira NGN1,000 = US$2.75). Data analysis was performed with content analysis, descriptive statistics and correlation analysis using the Pearson correlation coefficient.

Factors for success in studied cases

Prior to 2010, the dominant farming systems among the 150 interviewed farmers were crop rotation (56 per cent of farmers) and intercropping (43 per cent) (Table 4.2). The rotation systems allowed households to leave some fields under fallow for one year. After 2010, one-third used crop rotations, while two-thirds of the farmers engaged in intercropping. The differences between women and men increased after 2010. Women continued using crop rotation, while men shifted to intercropping and continuous cropping (Table 4.2).

By 2010, less agricultural land was available due to urban encroachment. Farmers shifted from rotational cultivation with some fields under fallow to continuous cultivation with intercropping with a short idle period of three months. Similar trends were found in Cross Rivers State after 2013, where 34 per cent of the farmers engaged in crop rotation, and 66 per cent practised continuous cultivation (Yaro *et al.* 2014).

Integrated farming with livestock was uncommon in both periods. Most farmers in this study said that rising food prices were the primary driver

Table 4.2 Peri-urban multiple-cropping systems in Port Harcourt

Farming system	Before 2010 per cent (women/men)		After 2010 per cent (women/men)	
Rotation, annual crops in sequence (including fallow)	56	(29/27)	31	(28/3)
Intercropping, continuous cropping	43	(21/23)	66	(21/45)
Integrated crops and livestock production	1	(1/0)	3	(1/2)

Source: Author's field data (2017).

Note
$n = 150$.

for peri-urban farming. Land scarcity was the main driver for adopting more permanent multiple-cropping systems (78 per cent) in use after 2010. The growing urban population and demand for food provided opportunities for commercial production of staple crops, including cassava, and 70 per cent of the survey respondents had assumed more intensive and continuous multiple-cropping systems.

Cassava is planted during four periods in Nigeria, depending on the agroecological zone (Ajayi 2014) and depending on the variety, it is ready to harvest in six months to three years. In Port Harcourt, the planting period is timed with the onset of the rainy season in March (Table 4.3). The long maturity of cassava lends it to intercropping during the first four months with a variety of crops, such as potato, vegetables and maize. Maize is the only crop grown twice, the first time intercropped with cassava, while the cassava gives too much shade for the second maize crop, which then is grown as monoculture or intercropped with shorter crops. Continuous cropping with fruits may include plantain, pawpaw (*Carica papaya*), soursop (*Annona muricate*) as well as some sporadic oil palm fruits (*Elaeis Guineensis*), and coconut (*Cocos nucifera*).

Cassava yields depend on spacing. The typical spacing of one-by-one metre (10,000 stems per hectare) still allows for intercropping. Estimates vary from 500 to 40,000 plants per hectare depending on variety, branching type, leaf shape, soil fertility status and agroecological zone. The intercropped plants may also affect cassava yields. Studies show that intercropping with legumes or cowpeas did not affect the yield compared to monoculture cassava, hence a higher Land Equivalent Ratio was seen with intercropping. The greatest economic return was seen with a cassava density of 10,000 plants per hectare and cowpea at 80,000 plants per hectare (Njuko and Muoneke 2008). Cassava yields vary across the country, and the low national averages of 8.8 to 10.3 tonnes per hectare in recent years (Figure 4.1) contrast with some earlier studies showing yields in the Nigerian rain forest belt, where Port Harcourt is located, of 15 tonnes per hectare and ten tonnes per hectare in the dry savanna (Oyekanmi and Okeleye 2007). The author's own unpublished interviews with 90 households in the area in 2011 found that the average cassava yield was 10 ± 3 tonnes per hectare.

Based on farmers' assessments, these cassava-based peri-urban systems in Port Harcourt can be understood as multifunctional. In addition to contributing to food and income, these intercropping and continuous cropping systems with palm trees, coconuts and small trees offer a lush green landscape that farmers consider aesthetically pleasing (illustrated by Figure 4.2). Furthermore, farmers said that they provide shade for microorganisms, insects and small wildlife (ecological functions) and enable preservation of traditional cassava landraces (cultural values, traditional menus and food security). Another benefit of multiple-cropping systems, according to the surveyed farmers, was that they enabled more efficient use of

Table 4.3 Farming calendar for the crops in the cassava-based systems in Port Harcourt

Jan.	Feb.	Mar.	Apr.	May	June	July	Aug.	Sep.	Oct.	Nov.	Dec.
		RAINY SEASON									
		CASSAVA (harvest up to 3 years)									
			MAIZE					MAIZE			
					PLANTAIN						
			SWEET POTATO, YAM								
			VEGETABLES								
		FRUIT TREES									

Source: Author's field data 2017.

Figure 4.2 Peri-urban multiple-cropping system with cassava intercropped with yam, and maize in Port Harcourt.

Photo credit: Onoja 2017.

small pieces of land, compared to cassava monoculture. The most preferred ecosystem functions were shade and landscape beautification (80 per cent of the interviewed farmers), green manure (70 per cent) and lower yield loss due to adverse weather and pest infestation compared to monoculture (65 per cent).

The main components of peri-urban multiple-cropping systems in Port Harcourt are illustrated in Figure 4.2, and Tables 4.3 and 4.4. The combinations of cassava-based systems and their productivity indices – gross margins and cassava yields – are presented in Table 4.4. The results show that the mix of cassava, vegetable, plantain and yam had the best outcome for cassava, both in terms of yields and gross margin for both women and men, on average 17.4 tonnes per hectare and US$420. This was on par with gross margins of US$401 in Akure (Oduntan *et al.* 2012), but the yields are considerably higher than those recorded in Port Harcourt by

Table 4.4 Gross margin of cassava in peri-urban multiple-cropping systems in households headed by men (M) and women (W) in Port Harcourt

Cassava-based system	Gender of household head	n	Farm size (hectares)	Cassava yield (tonnes per hectare)	Annual gross margin (US$ per hectare)
Cassava, plantain	M	9	0.7	13.1	229
	W	14	0.5	12.8	177
Cassava, maize	M	11	2.4	10.6	264
	W	14	2.2	9.5	248
Cassava, yam, maize	M	11	2.5	9.9	273
	W	11	2.5	10.6	269
Cassava, fruits, maize, potatoes	M	13	2.1	10.9	416
	W	12	1.2	10.5	243
Cassava, maize, vegetables	M	16	2.3	13.2	428
	W	11	1.8	14.3	325
Cassava, vegetables, plantain, yam	M	15	2.3	17.8	439
	W	13	2.1	16.9	396

Source: Author's field data 2017.

Notes
For details, see Figure 4.3.
$n=150$.

Oyekanmi and Okeleye (2007) and by the author (unpublished) in 2011. Furthermore, all the cassava mixes rendered gross margins per hectare (for cassava tubers only) between US$197 and 419, which is significantly higher than the US$36 recorded in Lagos (Aminu and Okeowo 2016) and US$102 per hectare in Ogun State (Bamiro *et al.* 2012). The comparatively high-yield and economic returns are in line with other findings from Delta state, where the Land Equivalent Ratio for cassava with yam, maize and vegetables reached 1.94 and returned a higher income, compared to cassava monoculture (Chukwuji 2008). Similar multiple benefits were recorded by Oguru *et al.* (2008) and Bamiro *et al.* (2012).

Interestingly, the results show that women-headed households consistently have smaller farms and lower net annual returns from cassava than men-headed farms, for all the cassava-based practices, even for the two practices for which women have higher yields (Figure 4.3). Some likely explanations for these differences are that women often rent land while men own it, and women have less available labour which creates a vicious cycle of a larger share of funds spent on expenses, with less capital for investments. These circumstances can influence the quality of the land and hence the yield. Women's access to land is still mediated via patriarchal systems where land is passed on to male descendants, despite the 1978 Land Use Act, through which all land was nationalized and authorized by the State Governor. Enwelu *et al.* (2014) suggested that the conditions required for land tenure are beyond the financial capacity of many women.

The annual gross margin for cassava ranged between US$343 and 1,845 per year for individual households (Table 4.4). This translates into household benefits, covering at least 45 per cent of an average urban household's expenses for food among peri-urban farmers. At least 55 per cent of the interviewed peri-urban farmers used the incomes to pay school fees, 43 per cent paid health bills, and 35 per cent bought furniture, fuelwood and animal fodder that they otherwise may not have been able to afford. Over 90 per cent said these benefits were attained after shifting to cassava-based multiple-cropping systems.

Drivers

The changes in peri-urban land use in Port Harcourt were primarily driven by the farmers themselves in response to changes in demand, markets and available land, which is what Yaro *et al.* (2014) also found. Compared to rural areas, peri-urban farmers have a number of advantages. In Port Harcourt, the distance between farmland or home gardens and markets is up to four kilometres, while it can range up to 25 kilometres in the countryside (compare with Shomkegh Chapter 2). This allows peri-urban farmers to sell their produce fast, reducing costs and risks associated with storage. As few farmers have vehicles, less valuable time is lost transporting products.

Limitations

With less than 5 per cent of the interviewed peri-urban farmers having access to agricultural extension services, the results show that these practices can spread autonomously. However, the needs for further improvements of these farming systems resemble those raised over a decade ago (Allison-Oguru et al. 2006), as nearly all respondents in 2017 still struggled with pests, diseases and weeds. Common cassava pests in the area include insects (especially cassava mealybug, cassava green mite and white flies) and rodents, particularly cane rats (*Thyronomys swinderianus*) and squirrels (for example, Geoffroy's ground squirrel *Xerus erythropus*).

The scarcity of urban and peri-urban land, resulting in the use of fragmented land holdings, was perceived by 78 per cent of the farmers to be a limiting factor for commercial scaling. Similar effects for urban expansion and a decrease in agricultural land are found in the South-South zone (Yaro et al. 2014). This study showed no significant correlation overall between farm size and cassava yield. However, the gender gap is significant both for farm size and yields (Figure 4.3), with almost only men in the positive anomaly (upper right quadrant) and almost only women in the negative anomaly (lower left quadrant).

Furthermore, Figure 4.3 shows that there are more diversified cassava-based systems, especially with fruit trees, on the larger farms. Extrapolating these results for smaller farm sizes and lower cassava productivity as proxies for fragmentation would suggest that interventions for urban and peri-urban agriculture need to focus on higher yields or a shift to higher-value crops, to help improve farmers' incomes.

In space-limited farming systems, such as peri-urban agriculture, farmers may choose to plant crops as densely as possible, which may lead to competition. Optimized spacing aims for one plant gaining benefits from the other or both plants benefitting mutually from each other to safe-guard against, for example, weather-related crop failures. Here we note that legumes were not included in the studied systems, although indicating positive yield benefits for both cassava and legumes in other studies. The degree of diversification may also generate different results. While this study focussed on cassava, the highest net returns from cassava were generated from the systems with three species, while cassava yields were slightly higher in the system with four components. This suggests that studies could focus on identifying combinations of trees and crops that offer mutual benefits and contribute to more resilient systems, with a more complex assessment of the total farm income and mutual ecological benefits from all crops in the system.

Training for farmers should include information technology for building climate resilience, such as weather information and early warning tools, and for accessing market information and farm inputs. One example is the computer programme Fertilizer Optimization Tool, which allows farmers

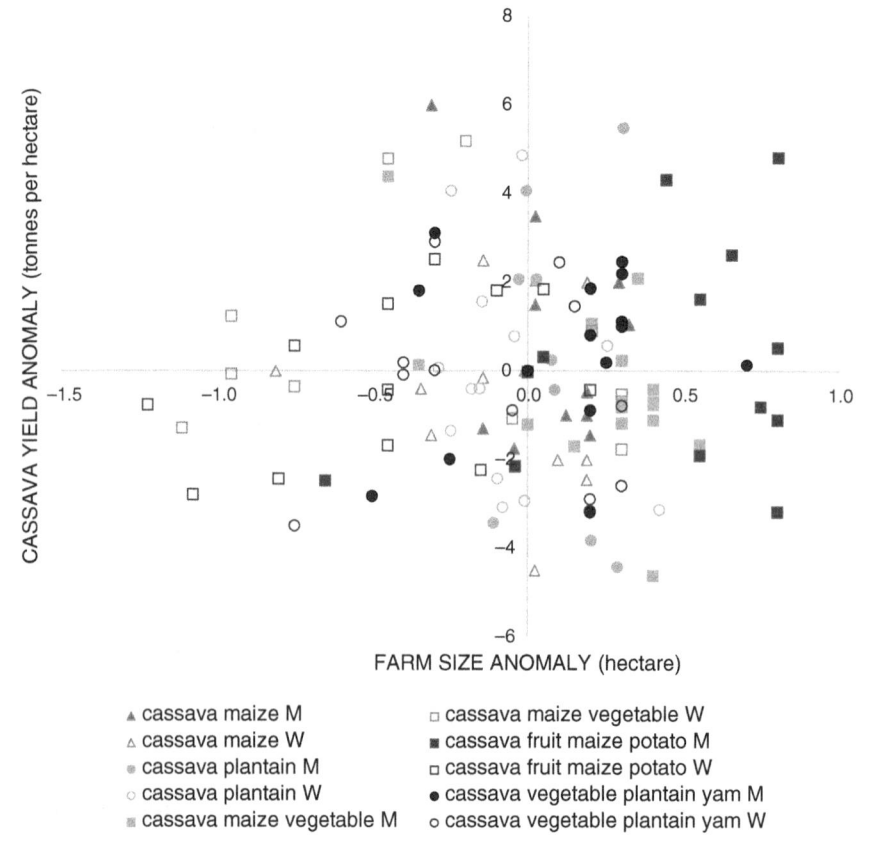

Figure 4.3 Relationship between farm size and cassava yield in multiple-cropping land use by gender (men – M solid symbols, women – W unfilled symbols; $n=150$).

Source: Author's field data 2017.

to enter the amount of money they can invest, field size, local cost of fertilizer and the market price of their crop and then calculates how much fertilizer to use to maximize the return-on-investment. The tool has been tested in 12 African countries, including Nigeria (TMP 2018).

Policy aspects

Despite Nigeria having signed the Sustainable Development Goals on poverty and hunger reduction (FAO and EU 2017) and the African Union Framework under the Malabo Declaration (AUC 2014), few policies seem to have created environments meant to enable peri-urban farmers to develop multiple-cropping systems for food security. The link between the global goals and the local reality is still to be bridged.

The current agenda of the Federal Ministry of Agriculture and Rural Development is a market-led agricultural transformation that focuses on developing agriculture value chains, agribusinesses and job creation through public-private partnerships to stimulate investments among farmer groups (FMARD 2018; Omolola 2015). In 2015, Nigeria launched the Agricultural Promotion Policy for 2016–2020, which sets targets for how domestic food security goals will be met through an agribusiness economic approach. The document prioritizes factoring in climate change and environmental sustainability by 'focusing policy instruments on the sustainability of the use of natural resources (land and soil, water and ecosystems) with the future generation in mind while increasing agricultural production, marketing and other human activities in the agricultural sector'. Moreover, the Nigerian National Policy on the Environment states that it focuses on 'abatement, remedial and restorative activities directed at: problems arising from industrial production processes; problems caused by excessive pressure of the population on the land and other resources; and problems due to rapid growth of urban centres' (FEPA 2018).

Unfortunately, the Agricultural Promotion Policy for 2016–2020 does not mention urban and peri-urban farming, nor how it can be developed in Nigeria (FMARD 2017). Nevertheless, training programmes that urban and peri-urban farmers can benefit from in Nigeria are being offered by the Agricultural and Rural Management Training Institute. The programmes cover a wide range of topics on planning and management of agricultural and rural development, including agricultural finance and rural credit, entrepreneurship and small/medium enterprise development, gender and youth, agricultural research and extension, and rural infrastructure (ARMTI 2018). Peri-urban farmers in Port Harcourt have yet to benefit from the programme. Policies are needed to reduce the cassava food deficit in the Niger Delta and nationally. According to Adedipe *et al.* (2010) as cited in Uche *et al.* (2016), the demand and supply deficit for cassava in the Niger Delta region is 9.5 tonnes annually. Efforts to enhance access to land for women-headed households and strengthen women in farm-business development in Port Harcourt could bolster opportunities to commercialize cassava-based cropping systems in the region.

Lessons learned from the case

These cassava-based multiple-cropping systems are economically beneficial farming systems that to some extent buffer shorter periods of food insecurity, diversify diets and improve incomes, with observed environmental benefits, such as green manure and pest control, compared to monocultures of cassava. Causes of declines in national level cassava yields (Figure 4.1) need to be clearly understood and compared with trends of cassava yields in multifunctional systems. Although these farming systems were largely driven by external factors, without support from national agricultural policies, the relative advantages and disadvantages of remaining the world's largest cassava producer without making higher-yielding varieties available should be further studied.

A detailed understanding of the gender differences is also needed, especially in cases like the one presented here, where most of the farm systems are headed by women. The persisting poor yields and lack of extension services suggest that these farming systems still underperform. A better understanding of the contexts in which high-yielding cassava varieties are suitable in multiple-cropping systems could benefit smaller farms. For example, none of the practices in this study used leguminous crops, while other studies have demonstrated yield benefits. This study showed that at the household level, important indirect benefits can be seen in how farmers invest their gains, namely in more diverse food and children's education.

Scaling would require a consorted policy effort in support of extension advisory services and providing access to farm credits or village funds. Community approaches can also serve to reduce the use of pesticides, not only among rural but also urban farmers. Meeting peri-urban farmers' needs calls for a closer collaboration among agriculture planners, universities, and national research and extension institutes to develop training and advisory services for farmers.

References

Ajayi JO. 2014. Comparative economic study of mixed and sole cassava cropping systems in Nigeria. *Agris on-line Papers in Economics and Informatics* 6(4):15–23. http://purl.umn.edu/196522.

Allison-Oguru EA, Igben MS, Chukwuigwe EC. 2006. Revenue maximising combination of crop enterprises in Bayelsa State of Nigeria: a linear programming application. Research Notes. *Indian Journal of Agricultural Economics* 61(4):667–676. https://ageconsearch.umn.edu/bitstream/204495/2/06-Allison.pdf.

Aminu FO, Okeowo TA. 2016. Economic analysis of cassava mixed farming enterprises in Epe Local Government Area, Lagos State, Nigeria. *Applied Tropical Agriculture* 21(3):122–130.

[ARMTI] Agricultural and Rural Management Training Institute. 2018. Brief on ARMTI (online). ARMTI. Jimba-Oja, Nigeria. https://armti.gov.ng/brief-on-armti/.

[AUC] African Union Commission. 2014. Malabo declaration on accelerated agricultural growth and transformation for shared prosperity and improved livelihoods. African Union Commission. Addis Ababa. 6p. www.resakss.org/sites/default/files/Malabo%20Declaration%20on%20Agriculture_2014_11%2026-.pdf.

Bakare AS. 2012. Measuring the Income Inequality in Nigeria: the Lorenz Curve and Gini Co-efficient Approach. *American Journal of Economics* 2(1):47–52. doi:10.5923/j.economics.20120201.06.

Bamiro OM, Afolabi A, Daramola F. 2012. Enterprise combinations in cassava-based food crop farming system in Nigeria: evidence from Ogun State. *Greener Journal of Agricultural Sciences* 2(1):13–20.

Chukwuji OC. 2008. Comparative analysis of enterprise combination costs and returns in cassava-based food crop farming systems in Delta State, Nigeria. Asian Research Publishing Network. *ARPN Journal of Agricultural and Biological Science* 3(4):27–32.

De Souza AP, Massenburg LN, Jaiswal D, Cheng S, Shekar R, Long SP. 2016. Rooting for cassava: insights into photosynthesis and associated physiology as a route to improve yield potential. *New Phytologist* 213(1):50–65. https://doi.org/10.1111/nph.14250.

Enwelu IA, Morah UL, Dimelu MU, Ezeano CI. 2014. Women farmers' access and use of land for agriculture in selected communities of Anambra State, Nigeria. *Mediterranean Journal of Social Sciences* 5(26):37–43. doi:10.5901/mjss.2014.v5n26p37.

[FAO] Food and Agriculture Organization of the United Nations. 2013. Save and grow. Cassava – a guide to sustainable production intensification. FAO. Rome. 142p. www.fao.org/3/a-i3278e.pdf.

[FAO] Food and Agriculture Organization of the United Nations. 2018*a*. Food Outlook – Biannual report on global food markets – November 2018. FAO. Rome. 104p. www.fao.org/3/CA2320EN/ca2320en.pdf.

[FAO] Food and Agriculture Organization of the United Nations. 2018*b*. FAO in Nigeria: Nigeria at a glance (online). FAO. Abuja. www.fao.org/nigeria/fao-in-nigeria/nigeria-at-a-glance/en/.

[FAO and EU] Food and Agriculture Organization of the United Nations and European Union. 2017. Strengthening sector policies for better food security and nutrition results: climate change. Policy Guidance Note 5. FAO. Rome. 38p. www.fao.org/3/a-i7217e.pdf.

FAOSTAT. 2018. Land use. The FAOSTAT land use domain (online). Food and Agriculture Organization of the United Nations. Rome. www.fao.org/faostat/en/#data/RL.

[FEPA] Federal Environmental Protection Agency. 2018. National Policy on the Environment. The Presidency and the UNDP supported Environment and Natural Resources Management Programme for Nigeria (NIR\C3). Abuja. 158p.

[FMARD] Federal Ministry of Agriculture and Rural Development. 2017. The Green Alternative (Nigeria's Agriculture Promotion Policy 2016–2020). Building on the successes of the ATA, closing key gaps. Policy and strategy document. FMARD. Abuja. 59p. https://fscluster.org/sites/default/files/documents/2016-nigeria-agric-sector-policy-roadmap_june-15-2016_final1.pdf.

[FMARD] Federal Ministry of Agriculture and Rural Development. 2018. Federal Ministry of Agriculture and Rural Development: What We Do. A Web-based Profile of FMARD. https://fmard.gov.ng/what-we-do-2/.

[IFAD and FAO] International Fund for Agricultural Development and Food and Agricultural Organization of the United Nations. 2005. Cassava development in Nigeria. A country case study towards a global strategy for Cassava development. In proceedings of *The Validation Forum on The Global Cassava Development Strategy* 2. IFAD and FAO. Rome. www.fao.org/3/a-a0154e.pdf.

Kahan D. 2013. Economics for farm management extension. Food and Agriculture Organization of the United Nations. Rome. 97p. www.fao.org/3/a-i3228e.pdf.

Mander Ü, Helming K, Wiggering H. 2007. Multifunctional land use: meeting future demands for landscape goods and services. *In*: Mander Ü, Wiggering H, Helming K. (eds). *Multifunctional Land Use*. Springer, Berlin, Heidelberg. P 1–13. http://link.springer.com/book/10.1007%2F978-3-540-36763-5.

[MBNP] Ministry of Budget and National Planning. 2016. National Policy on Food and Nutrition in Nigeria. Ministry of Budget and National Planning. Abuja, Nigeria. 41p. https://nigeria.savethechildren.net/sites/nigeria.savethechildren.net/files/library/NPFN%20manual%20design%20%20v13.pdf.

Mbow C, Noordwijk MV, Luedeling E, Neufeldt H, Minang PA, Kowero G. 2014. Agroforestry solutions to address food security and climate change challenges in Africa. *Current Opinion in Environmental Sustainability* 6:61–67. https://doi.org/10.1016/j.cosust.2013.10.014.

Métivier C. 2015. Violence and displacement in Northern Nigeria. *In*: Gemene F, Zickgraf C, Ionesco D (eds). State of Environmental Migration 2015 – A review of 2014. International Organization on Migration. Geneva. P 31–48. https://publications.iom.int/system/files/pdf/state_environmental_migration_2014_0_0.pdf.

[NBS] National Bureau of Statistics. 2018. National Bureau of Statistics: Demographic Statistics Bulletin. National Bureau of Statistics. Abuja, Nigeria. 26p. http://nigerianstat.gov.ng/download/775.

Njuko DN, Muoneke CO. 2008. Effect of cowpea planting density on growth, yield and productivity of component crops in cowpea/cassava intercropping system. *Journal of Tropical Agriculture, Food, Environment and Extension* 7(2):106–113. www.ajol.info/index.php/as/article/view/1591/513.

Nlerum F, Wechie E. 2018. Greater Port Harcourt City Urbanization Project and its socio-economic effect on affected farming communities in Rivers State. *Journal of Sociological Research* 9(1):91–101.

Oduntan O, Amos TT, Oseni JO. 2012. Efficiency and profitability of small-scale cassava production in Akure area of Ondo State, Nigeria. School of Agriculture Technology, The Federal University of Technology, Akure, Nigeria. *Applied Tropical Agriculture*. P 55–61. www.futa.edu.ng/journal/home/downloadpdf/paper_3_1508317286.

Oguru EA, Igben MS, Yahuwa TAT. 2008. Relative profitability of sole and mixed cropping enterprises in the central Niger Delta of Nigeria. *Indian Journal of Agricultural Research* 42(4):298–302. http://arccjournals.com/uploads/articles/ijar2424011.pdf.

Omolola AS. 2015. The political economy of food price policy in Nigeria. *In*: Pinstrup-Andersen (ed.) *Food Price Policy in an Era of Market Instability: A Political Economy Analysis*. Oxford Scholarship Online. United Nations University – The World Institute for Development Economics Research. Helsinki. 37p. doi:10.1093/acprof:oso/9780198718574.003.0013.

Onoja AO, Ajie EN. 2015. Cropping Intensification and Profitability of Cassava-Based Farms in Rivers State, Nigeria. Poster presentation at the International Conference on Integrated Systems Research on 3–6 March 2015. International Institute of Tropical Agriculture. Ibadan, Nigeria. www.slideshare.net/humidtropics/cropping-intensification-and-profitability-of-cassava-based-farms-in-rivers-state-nigeria.

Oyekanmi AA, Okeleye KA. 2007. Cassava production systems across some agroecological zones in Southwest – Northwest axis of Nigeria. *Asian Journal of Plant Sciences* 6(1):158–162. https://scialert.net/fulltextmobile/?doi=ajps.2007.158.162.

Ropo OI, Ibraheem AA. 2017. Response of cassava and maize yield to varying spatial scales of rainfall and temperature scenarios in Port Harcourt. *Research Journal of Environmental Sciences* 11:137–142. doi:10.3923/rjes.2017.137.142.

[RSG] Rivers State Government 2018. Port-Harcourt, Capital of Rivers State: The Garden City of Nigeria. Online Profile of Rivers State. Port Harcourt. https://riversstate.gov.ng/overview/.

Samuelson PA, Nordhaus WD. 2005. *Economics*. Tata McGraw-Hill, New Delhi. 776p.

Satterthwaite D, McGranahan G, Tacoli C. 2010. Urbanization and its implications for food and farming. *Philosophical Transactions of the Royal Society B: Biological Sciences* 365(1554):2809–2820. http://doi.org/10.1098/rstb.2010.0136.

[TMP] The Montpellier Panel. 2013. Sustainable Intensification: A New Paradigm for African Agriculture. Agriculture for Impact. London. 36p. https://ag4impact.org/wp-content/uploads/2014/07/Montpellier-Panel-Report-2013-Sustainable-Intensification-A-New-Paradigm-for-African-Agriculture-1.pdf.

[TMP] The Montpellier Panel 2018. Ecological Intensification. Technical Brief. Agriculture for Impact. London. 94p. https://ag4impact.org/wp-content/uploads/2015/07/Technical-Brief-Ecological-30-July2.pdf.

Uche C, Ajie EN, Familusi lC. 2016. A survey on the agricultural prospects and potentials of the Niger Delta, in a post oil era. *Global Journal of Science Frontier Research: Agriculture and Veterinary* 16(8):25–30. https://globaljournals.org/GJSFR_Volume16/4-A-Survey-on-the-Agricultural.pdf.

[UNDESA] United Nations Department of Economic and Social Affairs. 2015. World Population Projected to Reach 9.7 billion by 2050 (online). DESA. New York. www.un.org/en/development/desa/news/population/2015-report.html.

[WB] World Bank. 2018. Gini index. PovcalNet. Development Research Group, World Bank Group, Washington. https://data.worldbank.org/indicator/SI.POV.GINI?locations=NG.

Yaro MAU, Okon AE, Etam EB. 2014. Agriculture in the Urban Fringe: analysis of Calabar and its environs, Cross River State, Nigeria. *American International Journal of Contemporary Research* 4(9):122–130. www.aijcrnet.com/journals/Vol_4_No_9_September_2014/15.pdf.

5 Integrated aquaculture
Balancing food production systems and livelihoods in Kenya

Geraldine K Matolla

Current status of aquaculture

Global per-capita fish consumption has doubled since the 1960s to over 20 kilogrammes per year (FAO 2016). In the developing world, fish consumption increased more than ten-fold between 1981 and 1997 (De Silva *et al.* 2006). The development of freshwater aquaculture contributes to narrowing the gap between supply and demand (De Silva 2001) and is expected to reduce the pressure on global marine resources. In 2013 over 30 per cent of global fish stocks were fished at biologically unsustainable levels (FAO 2016). Aquaculture's contribution to the sector's food fish increased steadily from 7 per cent in 1974 to 39 per cent in 2004, and, in 2016, nearly half of global fish production came from aquaculture, 80 out of 171 million tonnes (FAO 2018). However, African aquaculture only made up 2.5 per cent of the total production in 2016, for reasons ranging from technical – a lack of breeds, feed and technical training – to non-technical – post-harvest losses, poor marketing infrastructure, and a lack of understanding among decision-makers (Chan *et al.* 2019).

Aquaculture has several socioeconomic benefits. First, fish and other aquatic products are rich in protein, essential amino acids, vitamins and minerals, hence important for reducing hunger and malnutrition. Africa's population is expected to double between 2018 and 2045, and urbanization is expected to increase the demand for fish (FAO 2018). Projections for the period 2015–2050s indicate that the fastest supply growth is likely to come from tilapia (*Oreochromis niloticus*), mullet (*Mugilidae spp*), catfish (*Clarias gariepinus*), and carp (*Cyprinus carpio*) (Chan *et al.* 2019). Second, aquaculture contributes to household incomes and offers employment opportunities (Subasinghe *et al.* 2009). FAO (2018) estimates that fisheries employ 5.5 million people in Africa, and aquaculture employs 303,500, with 70 per cent men, 11 per cent women, and the rest unspecified. However, with land and water becoming increasingly scarce across the continent, intensified models for integrating fish production with other farming activities hold promise for achieving food security objectives and improving livelihoods (De Silva 2001). Environmental impacts and the potential benefits for the poor from such aquaculture development remain debated globally (Chan et al. 2019; FAO 2018; Pant et al. 2014). Like land-based ecosystems, aquaculture provides ecosystem goods and services, such as food and feed, waste treatment, tourism and recreation (See Figure 1.3 on Land Equivalent Ratio in Simelton, Ostwald and Osiru Chapter 1). The historical and current aquaculture contexts are described below, followed by a presentation of two case studies of integrated aquaculture systems that provide several ecosystem functions in Kenya.

Aquaculture in Kenya

Historical development of aquaculture in Kenya

Kenyan aquaculture can be said to have evolved over three main eras.

Pre-independence era – recreation

Small-scale fish farming in Sub-Saharan African countries started effectively in the 1950s under the colonial powers (Nyandat 2005). In Kenya, the first fish farming developed from the introduction of sport fishing in the 1890s.

Post-independence era – donor driven

In the 1960s, rural fish farming became popular through the Kenyan government's 'Eat More Fish Campaign' (Aloo and Ngugi 2005), which promoted tilapia and catfish. The period between the early 1970s and early 1990s is regarded as the 'golden age' for aquaculture development, when donors supported partnerships between the national government and various public and non-government agencies. In the middle of the 1990s, this support declined as priorities shifted to other priorities, such as combatting HIV/AIDS and terrorism (Hecht 2006).

Between 1980 and 1996, aquaculture contributed less than 1 per cent of the total annual fish supply in Kenya (Neira et al. 2009). Kenyan aquaculture in the post-independence era was characterized by frequent shortages of supplies such as feeds and seed, insufficient extension services, and poor technical skills (Mwanja and Nyandat 2013). Non-governmental agencies such as the USAID-funded Aquaculture Collaborative Research Support Program were instrumental in the transfer and use of new fish-farming technologies, contributing to an increase in the annual aquaculture production from 1,500 to 2,500 kilogrammes per hectare between 1999 and 2005 (Quagrainie et al. 2010).

80 Geraldine K Matolla

Economic stimulus era – production boost

The Economic Stimulus Program (2009 to 2012) was initiated by the government to boost economic growth and lead Kenya out of recession. The initiative targeted business opportunities for aquaculture as a means of reducing poverty, particularly in rural areas. Farmers either adopted independent fish farms or integrated fish farming with crops and/or livestock (Nyandat 2005). Interventions included pond construction, stocking of fingerlings, and aquaculture advisory services. The Economic Stimulus Program led to a five-fold increase in aquaculture production between 2006 and 2014, with aquaculture representing 15 per cent of the total national fish production (KMFRI 2017). However, within four years of completion of the Program in 2012, aquaculture production dropped from 24,000 tonnes in 2014 to 16,000 tonnes in 2016 (Figure 5.1), as fish farms were abandoned due to the cessation of financial support. Similar outcomes and declines in food production were associated with discontinued government subsidies in Ethiopia and Nigeria.

Current status of aquaculture in Kenya

The two most common farmed fish species in Kenya are the Nile tilapia and African catfish, which account for about 75 and 21 per cent of the total aquaculture production by weight, respectively.

Aquaculture can be divided into three main categories: extensive, intensive and semi-intensive systems. Extensive culture in cages is mainly done

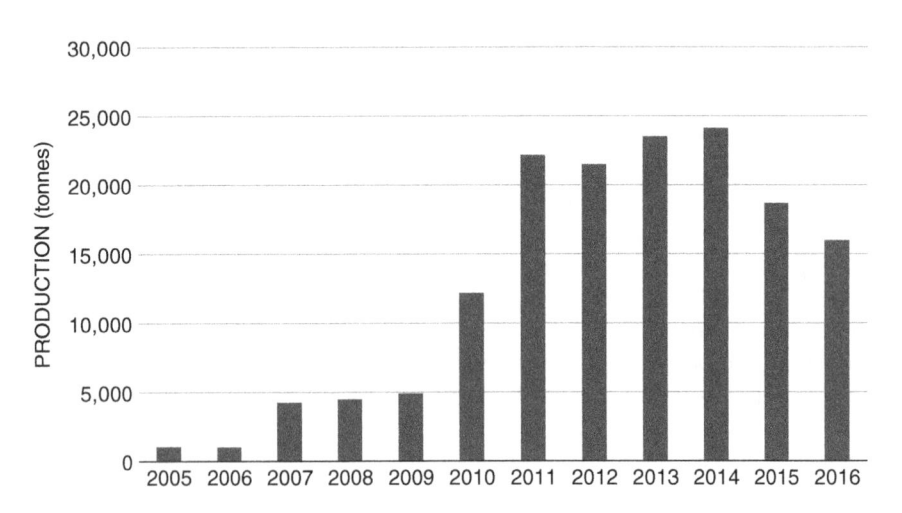

Figure 5.1 Freshwater aquaculture production between 2005 and 2016 in Kenya. The values of the Kenya Marine and Fisheries Research Institute (KMFRI) are similar to those reported in FAO, only 1100 tonnes higher than those reported by FAO in 2016.

Source: KMFRI 2017.

in lakes, rivers, dams and water reservoirs where the fish feed on organic matter that flows through the cages, with little to no input. As these systems depend on natural productivity and physical conditions, stocking densities are low, and yields range between 500 and 1,500 kilogrammes per hectare (NAFIS 2018). The main species are tilapia, catfish and common carp. The major advantages of extensive systems are low overhead costs for initial construction, as they can be set up in existing dams or lakes (Ngodhe et al. 2013), and few water quality deterioration problems, compared to intensive systems. They also allow for multiple water uses, such as fishing and recreational activities. The main drawbacks are that they are associated with excessive phytoplankton and algal growth, which reduces the levels of dissolved oxygen in the water and can cause the death of aquatic organisms, including fish (Ngodhe et al. 2013).

Intensive fish culture systems aim to achieve high production by keeping high stocking densities in more controlled environments, such as tanks, raceways and floating cages. Fish are produced by complementing or replacing the natural production with external feeding with aeration to ensure oxygen levels using mechanical filtration or bio-filtration. Intensive systems are often associated with high costs for energy and sophisticated equipment for monitoring water quality, feeding and aeration. Disadvantages with high stocking densities in intensive systems include problems associated with overcrowding, such as disease, water quality deterioration, and competition for food. The high start-up and operating costs are matched by the high production, which can range between 100,000 and 700,000 kilogrammes per hectare annually, depending on the technology (NAFIS 2018).

Semi-intensive systems blend extensive and intensive systems and account for more than 70 per cent of the total aquaculture production in Kenya (Nyandat 2005). Productivity reaches between 10,000 and 100,000 kilogrammes per hectare per year (NAFIS 2018). Ponds are fertilized using both organic and chemical fertilizers to enhance the natural productivity of fish feed, mainly algae and phytoplankton. Supplemental feed is usually made from locally available products, such as rice, wheat and maize bran.

Challenges to sustainable aquaculture development in Kenya

With the pressure for agricultural land increasing rapidly and investment capacity for most commercial intensive production systems beyond the scope of many smallholder farmers, the sustainable forms of integrated aquaculture intensification become a more affordable alternative for rural smallholders (FAO 2016; Little et al. 2016).

Aquaculture development in Kenya has been slow for several reasons. First, the sustainability of the value chain is endangered by weak regulation and policy guidelines. Second, current aquaculture policies focus predominantly on small-scale tilapia farming. With rising market prices for fish and strong public interventions, intensive production provides one option for profitable aquaculture business models (Nyandat 2005). However, the investment costs are beyond the financial means of most farmers, and the use of credit facilities for aquaculture expansion is relatively low compared to other agricultural production systems. This calls for concerted public and private interventions (Quagrainie et al. 2010). Third, the transition from semi- to more intensive fish farming is constrained by poor quality feed and seed, water quality management, and access to extension services. Affordable feed is one of the key challenges for Kenyan aquaculture. Studies show that feed accounts for 40 to 50 per cent of the total variable production costs (Munguti et al. 2014a). At the onset of the Economic Stimulus Program for aquaculture in 2009, the annual demand for fish feed was 14,000 tonnes, which quickly increased to 50,000 million tonnes per year in 2012. This prompted some dealers to produce and sell poor quality feed. An inadequate policy framework and the lack of standardized guidelines for the fish feed industry have led to substandard supplies of feed, contributing to losses on the fish production side and posing a risk to the sector's growth. Meanwhile, farmers use locally available ingredients, such as rice and maize bran and fish meal from dry fresh water shrimp and *Rastrineobola argentea*, also known as 'omena' (Liti et al. 2006; Munguti et al. 2014a).

The role of gender in Kenyan aquaculture

Kenya, like many African countries, has a patriarchal society, which largely influences gender roles, rural livelihoods and business opportunities. Gender refers to socially constructed norms, roles, and behaviours of men and women in a society. Gender determines social expectations on women and men, their access to assets and resources, decision-making and bargaining power, and control over benefits derived. Gender relations influence and intersect with cultural practices, domestic and social interactions, aspirations, and material livelihoods, and especially power relations and outcomes for well-being (Schumacher 2014).

Traditionally, fish trade used to require a small amount of start-up capital and offered income opportunities for women. However, the decline in the fish catch from Lake Victoria over the last two decades has profoundly changed livelihood opportunities, especially for women. With the transition to more intensive systems, women were becoming marginalized in both the trade and processing industry and are now instead turning to the input side of the value chain, by producing and selling fingerlings to fish farmers and for baitfish in Nile perch fishing (Williams et al. 2012). At the national level, estimates from 2011 show that fisheries and aquaculture employed 105,000 persons, of which one-third were women. However, out of the 7,840 persons who were employed in aquaculture, there were equal amounts of men and women. In an African context this ratio is high, only Zanzibar, Guinea and Mozambique had a larger share of women (de Graaf and Garibaldi 2014). The share of women employed in aquaculture and fisheries is likely underestimated, as many work part-time or temporarily.

However, the ownership of private ponds is a different story. Men own the majority of the land used for ponds, and fish ponds are regarded as household assets. In some cases, men may give some ponds to their wives, but remain the official owners of ponds. Women would only own ponds and the land when they are the head of the household. Typically, the income from fish sales is controlled by men in the households they head and by women in the households they head, but there are reports that male relatives have interfered to get access to this income in the case of households headed by women (Kiumbuku *et al.* 2013). Gender division of labour in the aquaculture sector in Kenya is evident. Men contribute to 79 per cent of pond preparation activities such as construction and stocking. Most of the pond management activities are carried out by women and children, such as feeding, fertilization, and predator control (Farm Africa 2016). Tending ponds takes relatively little time and can be combined with other homebased work. Women also add value to post-harvest production, such as filleting, salting, drying, canning and packaging. From this perspective, aquaculture is a women-dominated sector that can be integrated with other farming activities, such as recycling farm by-products and waste, multiplying benefits to home consumption and income (Jacobi 2013).

Although aquaculture has the potential to create livelihoods for many women, their opportunities are restrained by limited access to capital, land, water, training and markets. Studies in Kenya indicate that women suffer marketing challenges compared to men with respect to price information on products, and with respect to access and distance to markets (Farm Africa 2016). Some specific gender-targeted efforts have included linking women to market information systems, such as Farmed Fish Marketing Information System, which uses mobile technology to connect to credit, additional value chains, and training on aquaculture techniques (Akinpelu *et al.* 2013).

Awareness-raising activities about women's contributions to household food and nutrition security is needed to strengthen women's participation in aquaculture. This must be done through concerted efforts by gender champions, researchers, expert networks, and policy advocates and through the development of normative instruments for policy, projects, and programmes (Matolla 2015; Williams *et al.* 2012). Fish farm groups can reach many farmers for training and sharing information. Linkages with stakeholders such as feed manufacturers, extension services for fingerling producers, researchers, fish markets, and funding institutions and support for fish farmer groups/organizations with a specific focus on women's active and effective participation in group activities including leadership will be instrumental in improving gender equity in the Kenyan aquaculture sector.

Two aquaculture case studies

Selection of case studies

To understand how aquaculture-based land use affects food security and livelihoods in water-scarce areas, five farms in the arid and semi-arid regions of eastern Kenya were considered for case studies (see Figure 1.5 Simelton, Ostwald and Osiru Chapter 1). The sites were examined for promising evidence of:

1. water conservation initiatives in response to environmental and livelihood stress;
2. ecosystem services accrued from the land use, focusing on aquaculture-based systems;
3. contributions to livelihoods of local communities;
4. contributions to household food security;
5. multiple uses of land resources.

Only two farms recycled waste water from culture to crop production, thus recycling nutrients from fish waste. These two were selected to represent a semi-intensive and an intensive fish production system.

Musuu Horticultural and Fish Farm

Musuu Horticultural and Fish Farm is located in Tulia, Mutonguni in Kitui County, 170 kilometres south-east of Nairobi. The farm sits at 1532 metres above sea level in a semi-arid climate with temperatures ranging between 14 degrees Celsius (July to August) and 34 degrees Celsius (January to March) and annual total rainfall varying between 500 and 1,050 millimetres, on average 900 millimetres.

Kamuthanga Fish Farm

Kamuthanga Fish Farm is located in Machakos County, at the foot of Mua Hills, 82 kilometres from Nairobi. The farm is at 1,836 metres above sea level in semi-arid climatic conditions with temperatures ranging from 13 to 26 degrees Celsius. Long-term average annual rainfalls for two nearby stations measured 600 and 750 millimetres, ranging between 300 and 1,200 millimetres, with a dry spell between June and September (Huho 2017). Due to frequent rainfall failures, traditional farming systems with maize, beans, and cowpea have been abandoned for other ventures, including fish farming.

Figure 5.2 Musuu Farm has an open fish pond with water storage tanks that supply water for drip irrigation for tomatoes and with bamboo trees planted along river bed for preventing soil erosion.

Illustration by Simelton 2019.

Land-use changes at Kamuthanga Fish Farm

Kamuthanga Farm is a privately-owned farm that pioneered fish farming in the Machakos County and has become a national leader in producing quality fish. The fish production section now occupies about 10 per cent of the 4.9-hectare farm.

The farm started as a coffee estate in the 1970s. When global coffee prices plunged in the 1980s, many Kenyan farmers sought alternative farming activities, such as food crops, livestock, and real estate development. This changed after a visit by the proprietors of Kamuthanga Farm to

86 *Geraldine K Matolla*

Haller Park in Bamburi, Mombasa. The Haller nature park is a limestone quarry that has been rehabilitated from wasteland to wildlife park with a game sanctuary, snake park, crocodile farm, and an integrated fish farm facility since the 1980s. Although unfamiliar with fish farming, they identified a gap in fish supply and took up fish farming as a family business in 2004.

After constructing the first fish ponds, they discovered that the water supply from the on-site reservoir was insufficient. The farm invested KES2 million (US$20,000) in a borehole, which increased fish production from below 0.5 to two tonnes per month at that point. In collaboration with the Ministry of Agriculture, Livestock and Fisheries, the farm was identified as a potential fingerling production centre and received feed and technology support through the Economic Stimulus Program (2009 to 2012). In partnership with the Department of Fisheries, Kamuthanga became a fish-farming centre of excellence and served as a learning centre for farmers and students from primary schools to universities. Through local and foreign partnerships, the farm now invests in the most recent system for intensive fish production, with more effective breeding, rearing and marketing strategies. Imported quality fish feed from Egypt contributed to the annual production, approximately 100 tonnes of adult tilapia and over 100,000 fingerlings annually. The 100 tonnes translate into a production of 40.5 tonnes per hectare of adult tilapia, which is within the range of other intensive systems in the country.

The production levels are attributed to the fish-rearing technology (Figures 5.3a, b). Fish production begins with removal of the eggs from the mouth of the adult female tilapia and transfer to the hatchery, where nearly 200,000 eggs are hatched each month. Of these, 40,000 are grown to the adult stage, and the rest are sold to smallholder farmers at about KES10 (about US$0.10) per fingerling. The hatchery has a capacity of about one million fingerlings. Fingerlings are sorted regularly to maintain uniform size to control cannibalism. Within one month, fingerlings can be sold or transferred to a grow-out recirculating aquaculture system. By re-using water from fish production units, the recirculating aquaculture system minimizes water wastage. It is fitted with a bio-filter that removes nitrogenous waste from the fish and is supplemented with oxygen pumped into the water.

Due to overpopulation, stunted growth and non-uniform fish sizes associated with mixed sexes, male tilapia are preferred. To achieve all-male tilapia stocks, the farm administers male hormone to recently hatched fry (one week old), which leads to testicular tissue development in females that can then function reproductively as males. Such methyl testosterone treatment of fry is a standard technique used globally to produce male tilapia stocks, which grow faster and into a more uniform size than mixed or female tilapia. The quantities of hormones used at the farm are miniscule compared to the levels produced by men and women or consumed via

(a)

(b)

Figure 5.3 Kamuthanga Farm (a) Recirculating aquaculture system with (b) outdoor fish culture tanks close to the agriculture fields with bananas in the background and young tomato plants.

Photo credit: Matolla 2018.

growth hormones in meat and dairy products (Megbowon and Mojekwu 2014). Tilapia excrete the hormone and the levels drop to less than 1 per cent within four days after withdrawing treatment and are not detectable in adult tilapia, which are consumed at the earliest after five months. While there is greater concern about estrogens and their effects on wild fish, the environmental impacts of wastewater from tilapia production are understudied. Nevertheless, the impacts are considered minor to those associated with agricultural waste and sewage (Macintosh 2008).

Kamuthanga is one of the few fish farms in the region that have adopted the hormone-treatment technique, with a success rate of 95 per cent male stocks produced. After treatment, fish are kept inside a greenhouse structure that ensures light and (Figure 5.3a) constant temperatures of 28 to 30 degrees Celsius. This enables 24-hour schedules, which halves the time to reach marketable size (approximately 500 grammes) from eight to four months compared to in standard ponds where low night temperatures lead to daytime-only feeding. With further optimization of the system, the farm plans to double the annual fish production from 100 to 200–250 tonnes in 2019.

Factors for success in studied cases

Water-saving technologies

Musuu Farm integrates fish culture with horticulture, livestock and forestry. To overcome the challenge of water, the farm owner started adopting a series of rainwater harvesting practices in 2007. Rainwater and surface run-off were directed to underground wells and used for drip irrigation to minimize water loss (Figure 5.2). The bottoms of the fish ponds were covered with plastic liners to control water seepage. This was an initial low-cost solution; however, due to environmental concerns, the farm is now considering a concrete lining. To reduce evaporation, ponds are shaded with plastic covers and trees. Furthermore, the farm joined an afforestation programme to increase tree cover along river banks in and outside the farm and reduce soil erosion. As of 2018, over 2,000 bamboo trees have been planted, increasing the tree cover from nearly zero to 40 per cent. The bare rocky landscape now has almost 50 per cent tree cover.

Kamuthanga Fish Farm adopted the recirculating aquaculture system technology for highland-based aquaculture, which meant that less water was needed compared to standard pond culture systems. The proprietor estimates at least 10 per cent less water consumption. This shows new opportunities in water-scarce arid and semi-arid regions.

Figure 5.3b shows the outdoor tanks that channel water into banana plantations, vegetables and crop fields. Nutrient-loaded water from the aquaculture is used for irrigating and fertilizing horticultural crops and cereals. With a production of 40 tonnes of fish per hectare, Kamuthanga demonstrates how intensive aquaculture systems with recirculating water technology release land for other uses, such as, in this case, ecologically grown crops. The farm intends to install solar energy, which could further minimize operating costs and reduce greenhouse gas emissions.

Food and nutrition

At Musuu Farm, local fish and horticulture production, such as tomato, capsicum, kale, banana and pawpaw, have contributed to reducing hunger and malnutrition. Despite severe water shortages, annual fish production on the farm reached 2,000 kilogrammes per hectare in 2015 (Nyanzu, personal communication 2017). This can be compared with the national average fish production under semi-intensive systems which ranges between 1,000 and 2,500 kilogrammes per hectare annually (NAFIS 2018).

Demand for fish is high, with the average market price of table size fish (300 to 600 grammes) at KES500 per kilogramme (equal to US$5 per kilogramme). With a national fish supply of 300,000 tonnes and a demand of about 800,000 tonnes each year, aquaculture has a big gap to fill (Ndemo 2018). Rather simple technology, such as recirculating systems and constant water temperatures between 28 and 30 degrees Celsius, can increase fish production from 5–10 tonnes per hectare in systems with fluctuating temperatures (Mbuga 2002) to 40 tonnes as demonstrated in Kamuthanga.

Community benefits

Musuu Farm's activities and changes have inspired local communities to promote water harvesting, where they previously depended on shallow wells from which women and children had to carry water for domestic purposes for several kilometres. The farm offers casual job opportunities for, especially, youth and women who stay in the community rather than migrating to cities.

Kamuthanga Fish Farm is the largest aquaculture operation in Kenya, employing over 50 skilled workers, of which 25 are women and 25 men. The wages enable workers to buy fresh farm products and fish, which improves their diets. The incomes are also contributing to children's education. The women have formed a banking group, where members can take loans at low interest rates under the lean periods. The fish, from fingerlings to adult tilapia, is sold at markets in Machakos and Nairobi.

For its progressive work, Musuu Farm was awarded the Head of State Commendation in 2013 for contributing to food security and to social and environmental well-being. The reforestation initiatives also led to the farm being nominated by the Kitui County Government to spearhead local community forestry initiatives.

The Kamuthanga Fish Farm is certified for Hazard Analysis and Critical Control Points for food safety, which ensures that systematic preventive food safety measures are in place with regards to biological, chemical and physical hazards in the production. This has been a critical advantage in market expansion for the farm. The farm is also certified by EcoMark Africa, which aims to reduce environmental impacts of fish farming, with standards on the use of medicines, sourcing of fish feed, and good working conditions for employees. Furthermore, the farm offers training for farmers and governmental and non-governmental actors and has partnered with local and international organizations on research activities to enhance fish farming in the region.

Limitations

Water is the main limitation for both the case-study farms. Kamuthanga made considerable investments to minimize water use. While Musuu Farm struggled with low production due to low quality and quantity of fingerlings and feeds, Kamuthanga Farm became a centre of excellence in breeding through government support.

Like many small farms, Musuu had limited access to extension services. However, in this case farmer-to-farmer networks played an important role in transferring knowledge on fish-farming techniques. Some risks have been identified proactively that potentially could have negative impacts, if not checked. If entering wells or fish ponds, water polluted with agrochemicals from the horticulture could adversely affect fish production. The genetic vigour in wild fish stocks may be lost if farmed fish escape and reproduce with wild fish. Furthermore, disease outbreaks in the farmed fish may spread to wild fish populations through untreated effluents from the farm. Lastly, to avoid local food insecurity, sales to distant traders from Nairobi, Thika or Mwingi towns only take place when there is an excess of crops and fish. The distance itself deters most traders, but the integrated aquaculture-horticulture farmers have gained experience adapting to new market demands. Should transportation infrastructure improve, sales patterns could change.

As most credit facilities and financial institutions fail to recognize aquaculture businesses *as* businesses, in contrast to crop and livestock producers, both farms had no or restricted credits and loans. Kamuthanga had joined development partnerships and was in the process of establishing a second farm in 2017.

Policy aspects

The Kenyan government's economic agenda for the period 2018 to 2022 includes four pillars, the 'Big Four': manufacturing, food security and nutrition, universal health coverage and affordable housing. Kenya's 'blue economy' for conservation and sustainable use of aquatic resources is in line with the food security component of this policy. In this context, integrated systems with aquaculture and agriculture are considered a means to diversify land-based activities and achieve Sustainable Development Goals, such as sustainable and inclusive growth (UNDP 2018). However, to achieve the full potential of the aquaculture sector's contribution to food security, employment and poverty alleviation in environmentally sustainable ways, there is a need for a more concerted policy strategy (De Silva 2001).

- First, an integrated legal framework with clear guidelines for development of the aquaculture sector that includes the agencies for water resource management, agriculture and livestock, fisheries and natural resources and the National Environment Agency.

Overall, increased private sector involvement is crucial for investment in sustainable development of aquaculture in Kenya. However, this will require standardized frameworks for quality assurance and for making credit available. The government needs to decide its future roles in such public-private partnerships. The government can support a sustainable development of the aquaculture industry and aquatic resources of member states in the region through a number of interventions.

- Establish public-private sector partnerships for aquaculture growth, development of value chains, access to local and international markets, processing, and finances through local and foreign investment capital (Ridler and Hishamunda 2001).
- Craft policies that recognize small-scale aquaculture as business development models to ensure their access to sustainable credit from financial institutions, including loan guarantees. This requires training for financial institutions on fish farming as business models.
- Introduce incentives for making intensive aquaculture more commercially viable, including tax and duty-exempt status on imported ingredients for fish feed. This requires policies and guidelines on import regulations of inputs. Support to forming farmer cooperatives may drive political power and support a stronger negotiating position on prices.
- Establish standards and guidelines for hatchery and breeding programmes, with certification of farm operations for safe fish. Policies supporting public-private and local-international partnerships in horticulture and fish farming can be important drivers.
- Construct regulatory frameworks and reform institutions that can provide incentives to reduce the threat of externalities such as water pollution and overharvesting of groundwater. Moreover, a revised comprehensive policy is required, along with regulatory provisions on restricting fish movements, in order to prevent and control the spread of disease (Subasinghe 2005). This includes strengthened local capacity for managing fish health (Akoll and Mwanja 2012).

- Make investments in training programmes and capacity development in fish farming for technical and management levels from public and private development funds. Particularly, the government needs to reform and expand training, research, demonstration farms and extension services to meet the increasing needs and demands from fish farmers. Fish farming should be introduced in school curricula to create awareness about aquaculture technologies (Munguti *et al.* 2014*b*).

'The Blue Growth Initiative' is an ecosystem approach to fisheries and aquaculture that fits well with multifunctional land-use approaches. The initiative targets all stakeholders in the value chain and works through three phases. First, it puts in place enabling conditions, such as legislation, financial incentives and capacity development. Second, it transforms by testing different interventions. Finally, the initiative mainstreams actions to scale up public and private programmes and operations (FAO 2018). More documentation is needed to generate the evidence for larger-scale interventions. However, the two integrated aqua-and-agriculture farms presented here contribute to Kenya's aggregated lessons learned.

Lessons learned from the cases

The two case-study farms are located in areas with limited rainfall. They provide two models of investment level for achieving sustainable food security and livelihoods while enhancing environmental benefits. Musuu Farm demonstrates that an enterprise centred on contributing to the needs of local communities can succeed. This multifunctional approach to land use, with benefits spilling over to community members, serves as a model of success amid scarcity of resources.

Sustainable aquaculture depends on the promotion of viable aquaculture investments, including the potential opportunities available in integrated and cage systems. This requires connections between researchers, farmers, and governmental and non-governmental organizations for sharing information on various aspects of fish farming including feeds, water quality, seed availability and disease management. Research is needed to develop suitable fish-farming technologies that can be integrated with other land-use activities for sustainable water and land-use systems in arid and semi-arid areas. Water and nutrient recycling coupled with smart technologies can help overcome scarce land and water resources. Intensification of food production systems within a multifunctional land-use approach seems to be a viable option where pressure for agricultural land is increasing rapidly.

References

Akinpelu M, Akinbile L, Ayeloja A, Akinosho GA, George FOA, Adeyemi J, Adebisi G. 2013. Gender differentials in knowledge and utilization of Information and Communication Technologies (ICTs) among fish farmers in Ido Local Government Area of Oyo State, South-Western Nigeria. *Journal of Agricultural Economics and Development* 2(6):255–263.

Akoll P, Mwanja WW. 2012. Fish health status, research and management in East Africa: past and present. *African Journal of Aquatic Science* 37(2):117–129.

Aloo PO, Ngugi CC. 2005. Participatory approach the key to aquaculture development in Kenya. *7th Conference of the Aquaculture Association of Southern Africa*. Rhodes University, Grahamstown, South Africa.

Chan CY, Tran N, Pethiyagoda S, Crissman CC, Sulser TB, Phillips MJ. 2019. Prospects and challenges for fish for food security in Africa. Global Food Security 20:17–25. https://doi.org/10.1016/j.gfs.2018.12.002.

de Graaf G, Garibaldi L. 2014. The value of African fisheries. FAO Fisheries and Aquaculture Circular. No 1093. Rome, Food and Agriculture Organization of the United Nations. 76p. www.fao.org/3/a-i3917e.pdf.

De Silva SS. 2001. A global perspective of aquaculture in the new millennium. *In*: Subasinghe RP, Bueno P, Phillips MJ, Hough C, McGladdery SE, Arthur JR (eds.) *Technical Proceedings of the Conference on Aquaculture in the Third Millennium*, 20–25 February 2000 Bangkok, Thailand: NACA, Bangkok and FAO, Rome. P 431–459.

De Silva SS, Amarasinghe US, Nguyen TT. 2006. Better-practice approaches for culture-based fisheries development in Asia. *Australian Centre for International Agricultural Research*, ACIAR Monograph No 120. Canberra. 96p.

[FAO] Food and Agriculture Organization of the United Nations. 2016. The State of World Fisheries and Aquaculture 2016. Contributing to food security and nutrition for all. Food and Agriculture Organization of the United Nations. Rome. 200p. www.fao.org/3/a-i5555e.pdf.

[FAO] Food and Agriculture Organization of the United Nations. 2018. The State of World Fisheries and Aquaculture 2018. Meeting the sustainable development goals. Food and Agriculture Organization of the United Nations. Rome. Licence: CC BY-NC-SA 3.0 IGO 227p. www.fao.org/3/i9540en/I9540EN.pdf.

[FAO] Food and Agriculture Organization of the United Nations. 2019. Global Aquaculture Production 1950–2016 (online). Fisheries and Aquaculture Department. Rome. www.fao.org/fishery/statistics/global-aquaculture-production/query/en.

Farm Africa. 2016. Report for Kenya Market-Led Aquaculture Programme Gender Impact Study Submitted by ETC- East Africa to Farm Africa. 24p. www.farmafrica.org/downloads/resources/farm-africas-kmap-gender-impact-study.pdf.

Hecht T. 2006. Regional review of aquaculture development. For Sub-Saharan Africa – 2005. *FAO Fisheries Circular* No. 1017/4. Food and Agriculture Organization of the United Nations. Rome. 113p. www.fao.org/3/a-a0619e.pdf.

Huho J. 2017. An analysis of rainfall characteristics in Machakos County, Kenya. *Journal of Environmental Science, Toxicology and Food Technology (IOSR-JESTFT)* 11(4):64–72. www.iosrjournals.org/iosr-jestft/papers/vol. 11-issue%204/Version-2/H1104026472.pdf.

Jacobi N. 2013. Examining the potential of fish farming to improve the livelihoods of farmers in the Lake Victoria region, Kenya. Assessing impacts of Government Support. MSc thesis. University of Akureyri, Iceland.

Kiumbuku S, Mutinda J, Bernard J. 2013. Forms of gender inequalities in fish farming in Kwanza Division, trans-Nzoia County, Kenya. *Research on Humanities and Social Sciences* (online) 3(15) https://iiste.org/Journals/index.php/RHSS.

[KMFRI] Kenya Marine and Fisheries Research Institute. 2017. Kenya's Aquaculture Brief (2017): status, trends, challenges and future outlook. Kenya Marine and Fisheries Research Institute, Mombasa, Kenya. 12p. www.kmfri.co.ke/images/pdf/Kenya_Aquaculture_Brief_2017.pdf.

Liti DM, Mugo RM, Munguti JM, Waidbacher H. 2006. Growth and economic performance of Nile tilapia (*Oreochromis niloticus* L.) fed on three brans (maize, wheat and rice) in fertilized ponds. *Aquaculture Nutrition* 12(3):239–245.

Little DC, Newton RW, Beveridge MCM. 2016. Aquaculture: a rapidly growing and significant source of sustainable food? Status, transitions and potential. Proceedings of the Nutrition Society 75:274–286. doi:10.1017/S002966511 6000665.

Macintosh DJ. 2008. Risks associated with using Methyl Testosterone in Tilapia Farming. Sustainable Fisheries Partnership. Honolulu. 12p. www.sustainablefish.org/; http://media.sustainablefish.org/MT_WP.pdf.

Matolla GK. 2015. Sex for fish: the cruel fuel for HIV/AIDS in African fishing communities. *In*: Channa S, Porter M (eds). *Gender, Livelihood and Environment. How Women Manage Resources*. Orient Blackswan Publishers. New Delhi. Chapter 4.

Mbuga HM. 2002. The role of Fisheries Department in aquaculture development in Kenya. *Samaki News: a magazine of the Department of Fisheries of Kenya* 1(1):24–30. Fisheries Department, Ministry of Agriculture and Rural Development. Nairobi.

Megbowon I, Mojekwu TO. 2014. Tilapia Sex Reversal Using Methyl Testosterone (MT) and its effect on fish, man and environment. Faisalabad *Biotechnology* 13(5):213–216.

Munguti JM, Musa S, Orina PS, Kyule DN, Opiyo MA, Charo-Karisa H, Ogello EO. 2014*a*. An overview of current status of Kenyan fish feed industry and feed management practices, challenges and opportunities. *International Journal of Fisheries and Aquatic studies* 1(6):128–137.

Munguti J, Kim J, Ogello E. 2014*b*. An overview of Kenya aquaculture: current status, challenges and opportunities for future development. *Fisheries and Aquatic Sciences* 17(1):1–11.

Mwanja W, Nyandat B. 2013. Challenges and issues facing small-scale producers: perspective from Eastern Africa. *In*: Bondad-Reantaso MG, Subasinghe RP (eds.). Enhancing the contribution of small-scale aquaculture to food security, poverty alleviation and socio-economic development: report and proceedings of an expert workshop. *FAO Fisheries and Aquaculture Technical Paper* No 572. Rome. P 143–152. www.fao.org/3/a-i3118e.pdf.

[NAFIS] National Farmers Information Services. 2018. Aquaculture production systems: market information, livestock and agricultural enterprises and extension services. Nairobi. www.nafis.go.ke/?s=aquaculture.

Ndemo B. 2018. Unravelling the politics of fish in Kenya. Business Daily 24 October 2018 (online). www.businessdailyafrica.com/analysis/columnists/Unravelling-the-politics-of-fish-in-Kenya/4259356-4820824-mv4v6g/index.html.

Neira I, Engle C, Ngugi C. 2009. Economic risk analysis of tilapia production in Kenya. *Journal of Applied Aquaculture* 21(2):73–95.

Ngodhe SO, Raburu PO, Matolla GK, Orwa PO. 2013. Assessment of water quality, macroinvertebrate biomass and primary productivity of small water bodies for increased fish production in the Lake Victoria basin, Kenya. *Lakes and Reservoirs: Research and Management* 18:89–97.

Nyandat B. 2005. National Aquaculture Sector Overview. Kenya. National Aquaculture Sector Overview Fact Sheets. FAO Fisheries and Aquaculture Department. Rome. www.fao.org/fishery/countrysector/naso_kenya/en.

Nyanzu T. 2017. Proprietor at Musuu horticultural and fish farm. Personal communication.

Quagrainie KK, Ngugi C, Amisah S. 2010. Analysis of the use of credit facilities by small-scale fish. *Aquaculture International* 10:393–402.

Pant J, Barman BK, Murshed-E-Jahan K, Belton B, Beveridge M. 2014. Can aquaculture benefit the extreme poor? A case study of landless and socially marginalized Adivasi (ethnic) communities in Bangladesh. *Aquaculture* 418–419:1–10. doi:www.sciencedirect.com/science/article/pii/S0044848613004857.

Ridler N, Hishamunda N. 2001. Promotion of sustainable commercial aquaculture in sub-Saharan Africa. Volume 1. Policy Framework. *FAO Fisheries Technological Paper*. No. 408/1. FAO: Rome. 67p.

Schumacher KP. 2014. Gender relations in global agri-food value chains: a review. *Erde* 145(3):127–134. http://dx.doi.org/10.12854/erde-145-10.

Subasinghe R, Soto D, Jia J. 2009. Global aquaculture and its role in sustainable development. *Reviews in Aquaculture* 1:2–9.

Subasinghe R. 2005. Epidemiological approach to aquatic animal health management: opportunities and challenges for developing countries to increase aquatic production through aquaculture. *Preventive Veterinary Medicine* 67:117–124.

[UNDP] United Nations Development Program. 2018. A vehicle to articulate development issues and foster dialogue: leveraging the blue economy for inclusive and sustainable growth, Issue no 6, April 2018. 7p. www.ke.undp.org/content/dam/kenya/docs/UNDP%20Reports/Policy%20Brief%20%202018%20-%206-%20%20Blue%20Economy%20for%20Inclusive%20and%20Sustainable%20Growth.pdf.

Williams MJ, Agbayani R, Bhujel R, Bondad-Reantaso MG, Brugere C, Choo PS, Dhont J, Galmiche-Tejeda A, Ghulam K, Kusakabe K, Little D, Nandeesha MC, Sorgeloos P, Weeratunge N, Williams S, Xu, P. 2012. Sustaining aquaculture by developing human capacity and enhancing opportunities for women. *In*: Subasinghe RP, Arthur JR, Bartley DM, De Silva SS, Halwart M, Hishamunda N, Sorgeloos P (eds). Farming the Waters for People and Food. *Proceedings of the Global Conference on Aquaculture 2010*, Phuket, Thailand, 22–25 September 2010. Rome: Food and Agriculture Organization of the United Nations, Bangkok: Network of Aquaculture Centres in Asia-Pacific P 785–822. www.fao.org/3/i2734e/i2734e.pdf.

6 What integrated watershed management can deliver for the environment and livelihoods

The Ethiopian experience

Kassa Teka Belay

Integrated watershed management – current status

Many parts of Sub-Saharan Africa have high levels of extreme poverty, food insecurity and natural-resource degradation, with aggravating adverse impacts from climate change (Chisholm and Woldehanna 2012). As in many Sub-Saharan Africa countries, most rural households in northern Ethiopia depend on rainfed agriculture for their livelihoods (Teka *et al.* 2014). In the Tigray region in northern Ethiopia (see Figure 1.5 in Simelton, Ostwald and Osiru Chapter 1), most arable land is located on steep slopes where the rate of natural soil regeneration cannot keep up with the speed of soil erosion (Nyssen *et al.* 2009). The Ethiopian Highlands Reclamation Study (FAO 1986) estimated that over 50 per cent of the land area was significantly eroded, with a net annual soil loss from croplands appraised to about 100 tonnes per hectare. Land degradation, combined with population pressure and climate variability, constrain agricultural productivity (Deressa and Hassan 2009; Teka *et al.* 2013, 2014). Taken together, these challenges call for a societal shift towards a sustainable development model. Hence, the Ethiopian government and non-governmental organizations have implemented a range of integrated watershed management practices, from organic and inorganic fertilizers, and improved stress-tolerant seeds, to infrastructure for water harvesting and exclosures to keep land free from human and livestock interference. Watershed here refers to a sub-drainage area of a major river basin, whereas the integrated watershed management is a continuous adaptive process for managing human activities and ecosystems at the watershed scale (CCME 2016).

One concentrated intervention started in 1997, when the Tigray Bureau of Agriculture and Natural Resources with support from Irish Aid piloted an integrated community-based watershed management programme in five watersheds, each about 1,000 hectares and involving 500 households (Chisholm and Woldehanna 2012). The programme had six major objectives (GIZ 2015): (i) improve food and cash crop production for food security, (ii) improve soil and water conservation, soil fertility and land management using appropriate biological and physical measures and agricultural inputs, (iii) improve multiple water supplies for domestic, livestock and irrigation purposes, (iv) increase household incomes by diversifying agricultural and non-agricultural activities, (v) empower communities' sustainable development of local resources, and (vi) integrate community priorities by community-based health education, hygiene and sanitation, and savings, as well as to increase the status of women and girls in the target communities. To achieve these objectives, the region and its partners designed specific natural-resources management activities for agriculture and agro-business development. The Ethiopian government and the World Food Programme merged farmers' priorities with technical specifications for watershed management in a local-level participatory planning approach. In 2003, this was developed into the programme called Managing Environmental Resources to Enable Transitions towards more Sustainable Livelihoods (Tongul and Hobson 2013). Compared to previous land rehabilitation initiatives, a stronger emphasis was now placed on income-generating activities for households and integrated management at the watershed level.

There have also been government regulation interventions, such as in 2005 when the Tigray region adopted an integrated participatory watershed management strategy (Gebremichael and Waters-Bayer 2007), which includes the country-wide Productive Safety Net Programme, which aimed at rehabilitating natural resources, and building social infrastructure such as schools, health posts, farmer training centres and waste disposal facilities (WFP 2012). So far, few attempts have been made to follow up on the impacts of integrated watershed management technologies. A few studies, for example Hadush (2015), have reported on the outcomes of the integrated watershed management programme in a structured way, which makes these reports difficult to use as a regional baseline as evidence for policy makers. Furthermore, except for India, the integrated watershed management strategy is scarcely implemented in semi-arid areas, where people's livelihoods are particularly restricted by water deficiency. Therefore, this study aimed to gather information from secondary reports and conduct a field survey to initiate a baseline for documenting good multi-functional land-use practices in support of planning and scaling processes. This study documented the results from Tigray in a way that researchers, planners and decision-makers can access and use as a baseline for further research, as evidence for policy and practice that can serve to improve the development of sustainable community activities and to scale up efforts, in similar contexts.

Case study of integrated watershed management in Tigray

Data and methods

The literature review included over 30 published and unpublished reports from regional bureaus, district offices, non-governmental organizations, and research and academic institutes working in the Tigray region. A selection of references and key indicators is provided in Table 6.1. Additionally, a field observation of the reviewed watersheds was conducted to document context-specific performance information (Table 6.2).

Study area

The Tigray region covers 54,572 square kilometres in northern Ethiopia (Figure 6.1), located between 12°15′–14°50′N and 36°27′–39°59′E. The major watersheds considered in this review are provided in Table 6.2. The topography varies from 500 to 4,000 metres above sea level (Teka et al. 2014). Many soils are weakly developed resulting from ongoing erosion processes, such as regosols, cambisols, arenosols, xerosols and leptosols. The most dominant reference soil group is leptosol, which is a thin soil on hard rocks that covers about 75 per cent of the region (Zenebe et al. 2013). Deeper soils suitable for agriculture include luvisols and andosols with high nutrient content, and fluvisols located on alluvial plains. With irrigation, vertisols and calcisols are used for grain crops or grazing (Nyssen et al. 2008; Teka et al. 2015).

The climate in Tigray is predominantly semi-arid, characterized by sparse rainfall and frequent droughts, with the average annual rainfall varying from 200 millimetres in the north-eastern lowlands to 1,000 millimetres in the south-western highlands. In most of Tigray, about 75 per cent of the rainfall is confined to two months, July and August. The average annual night and day temperatures are 15 and 25 degrees Celsius, respectively, with the minimum and maximum recorded in December and May, respectively (Teka et al. 2014).

The population of Tigray is about 4.5 million, with the rural population making up 80 per cent. About 85 per cent of the Tigray population earn their living from agricultural activities. The main livelihoods are annual crops such as teff (*Eragrostis tef*), wheat (*Triticum aestivum*), maize (*Zea mays*), barley (*Hordeum vulgare*) and sorghum (*Sorghum bicolor*), and animal husbandry, such as cattle, goat, sheep, poultry and bee keeping, where practices are entirely based on traditional technology with animal traction (CSA 2008; Teka et al. 2014). The average population density is 84 residents per square kilometre (CSA 2008), giving a typical holding per household of less than one hectare (Teka and Haftu 2012).

Table 6.1 Literature review of selected references

Reference	Watershed name	Indicator: Land cover/Land use	Soil erosion	Irrigation and water access	Yields (crop, dairy)	Income	Other focus
AWS 2012	Abarah We-Atsbaha	x		x	x		Policy
Gebregziabher et al. 2016	Abarah We-Atsbaha			x			
Negusse et al. 2013	Abarah We-Atsbaha			x			
Debalkew 2014	Messebo				x	x	Gender
Gebremeskel 2018	Hintalo	x	x				Soil health
Hadush 2015	Mariam-Shewito	x		x	x	x	
Haregeweyn et al. 2012	Enabered	x	x				
Kirubel and Gebreyesus 2011	Medego	x	x				Gender
Smur 2017	Adikisandid, Mesanu, and Tsaedanaele			x		x	
Teka et al. under review	Gulle	x	x	x	x	x	
Yaebiyo et al. 2015	Sheka			x	x	x	

Table 6.2 Environmental change and social impact indicators

Category	Indicator	Units
Land degradation status and changes	Land use, land cover change (change in vegetation cover, farm- and pasture-land), Soil erosion status and change	Hectare, per cent Erosion status in tonnes per hectare per year and change in per cent
Livelihood/food and nutrition security status and change	Changes in irrigation and water development, compared to before integrated watershed management and with non-rehabilitated watersheds Changes in fruit, vegetable and grain crop productivity Changes in livestock production Changes in household income and expenditures	Irrigation water volume in cubic metre, depth (metre) and shallow water wells development (in number) Tonnes per hectare Litres of milk per cow Income (birr per household per year)

Table 6.3 Spatial distribution of studied watersheds in Tigray, with average elevation

Administrative zone	Watershed	Elevation (metres above sea level)	Area (hectares)
East	Hintalo	2,555	1,838
	Gulle	2,041	1,382
	Abraha Atsbaha	1,966	6,667
South-east	Messebo	2,264	2,141
Central	Mariam-Shewito	2,071	3,502
	Medego	2,054	1,090
	Enabered	2,002	1,208
	Sheka	1,744	594

Integrated watershed management interventions

Although rainfed practices dominate, the use of small-scale irrigation has been expanding over the last two decades (Teka *et al.* 2014). For example, following the drought in 2002, small-scale household rainwater harvesting ponds were introduced by the regional government (Teka *et al.* 2014). Moreover, Gebreyohannes *et al.* (2012) reported that between 1994 and 2003, 54 large dams, each with an average water storage capacity of 1.0 to 3.5 million cubic metres, were constructed.

The major integrated watershed management practices and technologies implemented in the Tigray region are: (i) physical soil and water conservation measures, such as soil bunds, deep trenches, hillside terraces, trench and check dams, where the embankment is made of soil and stone with a basin in the lower part, so-called fanya-juu (Hurni 1993), (ii) water harvesting methods, such as check-dam ponds, shallow ponds for domestic use and irrigation, percolation ponds and pits, and spring development, and (iii) biological soil and water conservation measures, for example, re/afforestation, exclosures, agroforestry, and organic and inorganic fertilizers for soil improvement (Hadush 2015). A survey by the Regional Bureau of Agriculture and Rural Development showed that between 1997 and 2015, integrated watershed management practices and technologies had been introduced on 510,000 hectares of land, and 1307 community watersheds developed in Tigray. Tigray was the first regional state in Ethiopia where soil and water conservation measures were implemented extensively through collective decision-making. For example, communities contributed 20 to 40 free labour days every year. For this work, the region won the Future Policy Award by the World Future Council and the United Nations Convention to Combat Desertification in 2017 (UNCCD 2017). According to estimates, men, women and children moved at least 90 million tonnes of soil and rock to restore the landscapes across one million hectares (UNCCD 2017).

Factors for success in studied cases

The integrated watershed management interventions have had positive impacts on environmental sustainability and livelihoods in the Tigray region. The improvements include land cover, soil health, water access and food security.

Physical changes from land cover change

Compared to 2006, Tigray landscapes are now greener (Figure 6.1) (Teka *et al.* under review; UNCCD 2017). A time series study at the Gulle watershed in 2015 indicated an expansion of grassland by 4 per cent and bushland by 9 per cent, while bareland declined by 16 per cent (Teka *et al.* under review). Furthermore, in the Hintalo watershed degraded grazing lands had re-greened into shrub and bush cover (Gebremeskel 2018).

Compared to the initial integrated watershed management implementation in 1997, the survival rate of tree seedlings planted by the community had improved by 35 per cent and by 21 per cent on private plantations by 2005, and the species diversity had increased by over 30 per cent (Kirubel and Gebreyesus 2011). In the Messebo watershed, following Shannon's diversity index, Debalkew (2014) counted 1208 plant species on rehabilitated hillsides compared to 269 plant species in neighbouring non-rehabilitated land

Figure 6.1 Land cover changes in the same area in the Hintalo watershed in 2006 (left) and 2016 (right).

Photo credit: Teka 2006, 2016.

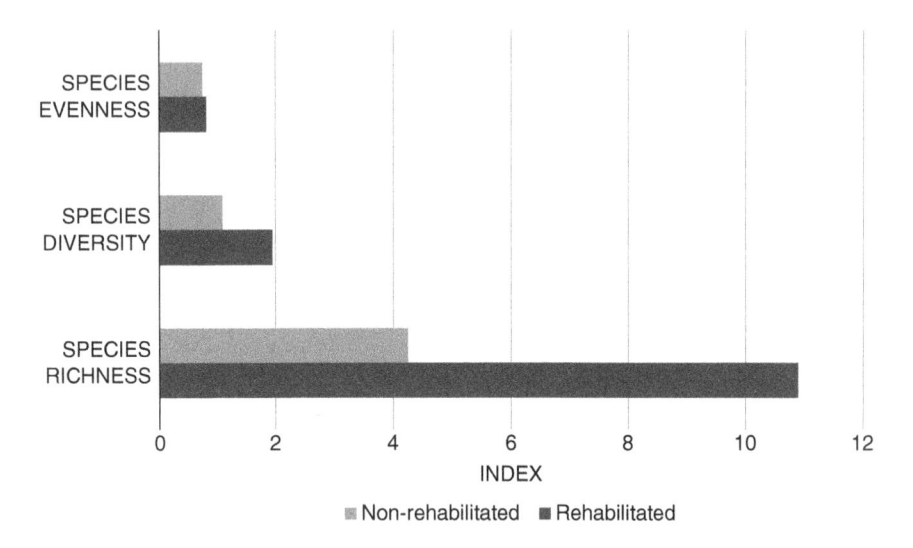

Figure 6.2 Species evenness, diversity and richness in rehabilitated and non-rehabilitated land.

Source: Adapted from Delbakew 2014.

(Figure 6.2). Species diversity, richness and evenness were significantly higher in watersheds with integrated watershed management compared to untreated ones. These increments were not due to changes in rainfall, but to the land restoration initiatives (UNCCD 2017).

Changes in soil quality

One of the direct benefits of re-greening watersheds was seen in soil quality. First, the litter-fall and roots from regenerated and planted trees increased the soil organic matter and enhanced biophysical processes, which improved the infiltration capacity of the soils and thus less water was lost as run-off and evaporation. The increased vegetation cover and surface roughness reduced soil erosion (Kirubel and Gebreyesus 2011; Teka *et al.* under review; Tongul and Hobson 2013). For example, in the Enabered watershed, surface run-off decreased by 27 per cent, from 7.92 million to 5.75 million cubic metres between 2004 and 2009 (Haregeweyn *et al.* 2012), while in Gulle watershed the average soil loss halved from 29 to 14 tonnes per hectare per year between 2002 and 2015 (Teka *et al.* under review). In the same year, the sediment concentration reduced from 30 grammes per litre before the intervention in 2002 to less than five grammes per litre after the intervention in 2015. In the Medego watershed, the average annual soil loss from sheet and rill erosion decreased from 117 tonnes per hectare before the interventions in 2004 to 12 tonnes per hectare in 2009 (Kirubel and Gebreyesus 2011). In the Hintalo district, four soil health indicators – soil organic matter, nitrogen, plant-available phosphorus and the amount of mycorrhiza – were compared in catchments with and without integrated watershed management, for two different geological origins (Figure 6.1; Gebremeskel 2018). Figure 6.3 shows that all soil quality parameters were higher in the watersheds with treatments. The

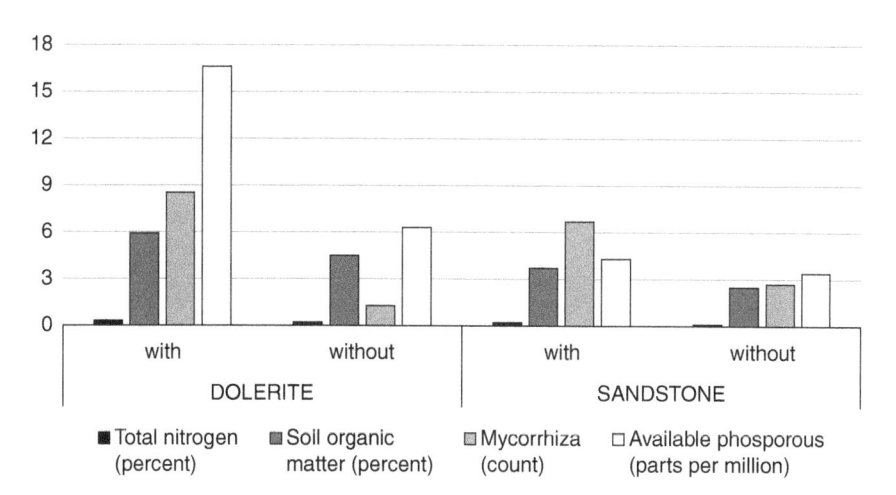

Figure 6.3 Averaged indicators of soil quality with and without integrated watershed management on dolerite and sandstone watersheds: total nitrogen (per cent), soil organic matter (per cent), mycorrhiza (count), available phosphorous (parts per million).

Source: Adapted from Gebremeskel 2018.

difference was significant for phosphorous on sandstone and for all four variables on dolerite, which is more responsive as the rock is less porous.

Changes in irrigation and domestic water availability

The integrated watershed management raised groundwater levels by up to five metres (Gebregziabher *et al.* 2016; Negusse *et al.* 2013). After the intervention, water was found at three metres' depth compared to eight, before. This improved groundwater recharge was associated with mixed root depths and permanent vegetation cover, which reduced surface run-off and instead improved soil moisture (Descheemaeker *et al.* 2009). According to the authors, the interaction effects of increased infiltration, more efficient water use for biomass production, and regenerated vegetation, resulted in up to 30 per cent of the annual rainfall percolating through the root-zone to groundwater recharge. The findings are interesting as there is limited evidence for the benefits of watershed management with tree cover in semi-arid areas and reluctance towards tree planting out of fear that this will lower groundwater tables (Ilstedt *et al.* 2016). The higher groundwater levels then allowed most of the previously rainfed agriculture on low-laying valley plains to be irrigated (Teka *et al.* under review). For instance, in the Sheka and Gulle watersheds, ten hectares outside the watersheds were adjoined through an irrigation canal, and 33 hectares of rainfed fields were converted into irrigated land (Yaebiyo *et al.* 2015; Teka *et al.* under review). Figure 6.4 shows irrigation water development at the Mariam-Shewito watershed.

Changes in food, feed and livelihoods

Several socioeconomic benefits were observed from the interventions in the watersheds. The new groundwater levels had direct impacts on daily life. Taking the Gulle watershed as an example, between 2002 and 2015 the walking distance between water points and homesteads reduced from on average 1.5 to 1.0 kilometre for 57 per cent of the sampled households ($n=269$). In this way, the national standard of one kilometre was met (ADF 2005). Furthermore, the daily domestic water consumption increased from 10 to 25 litres per person, which exceeds the national standard of 20 litres (ADF 2005). For comparison, these amounts are equal to 2–3 toilet flushes, where an average flush volume is 6 to 14 litres.

Furthermore, with more water available for irrigation, farmers were able to increase the farmland area and diversify the repertoire of crops. First, previously unproductive areas could be brought into production, and the cultivated area increased by 20 to 50 per cent (IWMI 2012; Teka *et al.* under review). Second, on the valley plains, farmers could grow a range of fruits, such as grafted orange (*Citrus sinensis*), lemon (*Citrus limon*), guava (*Psidium guajava*), avocado (*Persea Americana*), gesho (*Rhamnus prinoides*), papaya (*Carica papaya*), apple (*Malus domestica*) and coffee (*Coffea*

Figure 6.4 River water diversion at the Mariam-Shewito watershed, constructed between 2004 and 2006.

Illustration by Simelton 2019.

arabica), together with vegetables, such as onion (Bombay red), tomato (Roma VF), cabbage (Giant variety), hot pepper (Marko fana) and potato (*Solanum tuberosum*) (Hadush 2015; Teka *et al.* under review). Third, the average yields of grain crops such as wheat, teff, maize, barley and sorghum increased from 1.6 to 2.2 tonnes per hectare between 2002 and 2015. Teff and sorghum yields increased by over 60 per cent and maize by 27 per cent (Teka *et al.* under review; compare with conventional yield Figure 1.4 in Simelton, Ostwald and Osiru Chapter 1), attributed to the increased soil fertility and water availability in the treated watersheds.

Moreover, across Tigray, a three to four-fold increase in woody/tree biomass and forage production was reported (Tongul and Hobson 2013), which for example encouraged landless and young farmers to start honey bee production (Debalkew 2014, Teka *et al.* under review). In the Messebo watershed, 120 bee hives produced about 636 kilogrammes of honey annually, while there was no such activity in the neighbouring untreated watersheds (Debalkew 2014). A study in the Abraha We Atsbeha watershed (Chisholm and Woldehanna 2012) reported that honey production in the watershed has increased by 300 per cent over three years and incomes from vegetable and spice production tripled.

Moreover, in the Gulle, Sheka and Abraha We Atsbeha watersheds, the fodder production increased by 33 to 100 per cent between 2002 and 2015, where the improved feed resulted in local cows increasing production from 1.5 to 2.5 litres of milk per day (IWMI 2012; Teka *et al.* under review; Yaebiyo *et al.* 2015).

The watershed interventions coincided with the introduction of improved agricultural inputs, such as improved seeds and fertilizers (both organic and inorganic) and contributed to raising yields and farmers' incomes and diversifying diets (Debalkew 2014; Hadush 2015; Teka *et al.* under review; Yaebiyo *et al.* 2015). For example, 72 per cent of the survey respondents in the Gulle watershed were able to cover their annual expenditures in 2015, compared to about 50 per cent before the interventions. Another household survey in the Mariam-Shewito watershed indicated that the combined production of teff, wheat, maize and barley increased by 169 per cent from 1.3 in to 3.5 tonnes, with the average cash income increasing by 777 per cent, from ETB 1350 to 11,900 (US$50 to 420), within 18 years of the intervention (Hadush 2015). The change in the number of food-secure months was remarkable. In particular, the share of households who had food for less than six months reduced from about 40 per cent before the watershed implementation to 10 per cent afterwards (Figure 6.5).

Moreover, in the Messebo watershed, additional benefits from fodder, roof grass and bee keeping doubled the average household annual incomes from ETB 10,000 to 22,500 per year (approximately US$400 to 800), compared to before the integrated watershed management (Debalkew 2014).

Communities in most watersheds with interventions perceived several social benefits. For example, watersheds that implemented the interventions became popular for their rich sources of local herbal medicines for humans and animals. Women and children benefitted particularly from the irrigation facilities, which allowed them to spend less time fetching water and instead grow more vegetables and food crops. One study shows that women-headed households with irrigation were able to raise their incomes by 69 per cent, while households without irrigation did not see such increases (Smur 2017). The extra income was used to cover school, medical and other expenses and for buying additional food to diversify diets. Notably, the number of student dropouts declined by 34 per cent and youth job migration declined by 47 per cent (Teka *et al.* under review).

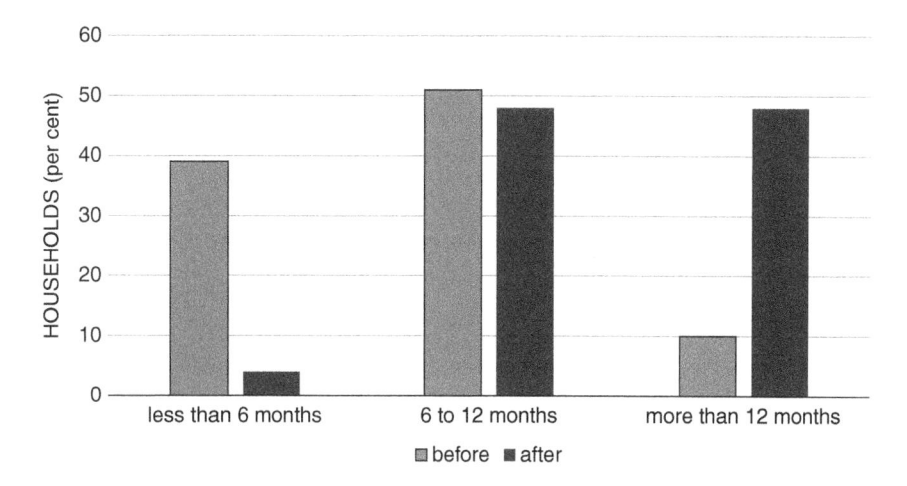

Figure 6.5 Number of months that households (*n* = 1,676) were able to cover their own food demand, before and after the integrated watershed management interventions in the Mariam-Shewito watershed.

Source: Adapted from Hadush 2015.

The shortened distance to water points and the provision of more fuel-efficient stoves contributed to healthier air quality, and freed up time, which also enabled women to participate in family and community decision-making and management. More than 90 per cent of the women respondents from the three villages Adikisandid, Mesanu and Tsaedanaele in Kilte-Awulaelo district, in eastern Tigray, said that their control and decision-making power over resources had improved. Furthermore, women's participation in community issues increased by 97 per cent due to the increase in income (Smur 2017).

At the farm level, the interventions generally enabled smallholders to diversify crop selection and use human and natural resources more efficiently, which allowed them to reduce and spread risk and converted losses into increased profitability (Hadush 2015). At the watershed level, the integrated watershed management approach was recognized by the government and development organizations for addressing interrelated problems of land degradation, low agricultural productivity, and food insecurity (Tesfaye *et al.* 2016).

Limitations

Despite the benefits, the sustainability and expansion of integrated watershed management interventions are challenged by at least seven factors in semi-arid Sub-Saharan Africa, and particularly in Ethiopia.

Low community participation

Farmers in some watersheds were reluctant to participate in the interventions when communities were not engaged in the planning process and local knowledge was not integrated in the proposed solutions. The planning processes focussed more on technical and physical activities, with less attention to the economic viability and social acceptance of the proposed interventions. This also resulted in communities being reluctant to take responsibility for the installed interventions (Chimdesa 2016).

Land and tenure security

Even though land certificates are provided to households, the certificate is only awarded for farmland while other land uses remain under state ownership. This restricted tenure limits the households' sense of ownership, and the sustainability of integrated watershed management practices and technologies remains questionable (Gorfu 2016).

High investment and maintenance costs

The investment costs for constructing fanya-juu bunds in the three watersheds presented here were estimated between US$29 and 87 per hectare, with annual maintenance costs between US$1.7 and 6.1 per hectare (Tesfaye *et al.* 2016). Communities would not be able to afford these costs without external support from government and non-government organizations. While major banks and micro-finance institutions provide loans to farmers for the purchase of improved seed, livestock and fertilizer, none of them offers loans for soil and water conservation maintenance. Hence, to maximize the benefits from integrated watershed management, further studies should investigate affordable conditions under which these institutions could expand loan services.

Weak institutional links

The level of coordination among researchers, extension centres and educational institutions was relatively poor, which affects the development and transfer of technologies from researchers to local experts and local communities, particularly farmers (Chimdesa 2016). Furthermore, when frequent restructuring of government institutions causes high staff-turnover rates, this leads to discontinued activities and limits the opportunity to expand promising watershed practices.

Incentives dependency

To compensate for the labour inputs of food-insecure rural households, the World Food Programme and the affiliated Safety Nets Programme provided cash and grain incentives most months during the year. These incentives are believed to influence the sustainability of the interventions when farmers end their involvement in the programme and their willingness to participate and work decreases. Conversely, the food-secure households may be less motivated to participate without any compensation, with such a considerable work load (Chimdesa 2016).

Frequent changes in technologies

As new technologies are introduced, they replace existing integrated watershed management technologies. Such a change may be initiated by the government without prior detailed study on the suitability of the new technology to the area. For example, the introduction of bench terraces in 2012 to create land for landless youth (EWAO 2013) replaced existing technologies, such as exclosures and stone terraces, without any prior impact assessment. Some farmers, therefore, lose confidence in the sustainability and effectiveness of new technologies, while studies from Kenyan drylands show that exclosures have indeed been beneficial for farm diversification while transforming from pastoral to livestock-based agropastoral systems (Nyberg *et al.* 2015).

Farmer preferences for short-term benefits

Since agriculture is the main occupation and means of livelihood for rural communities, farmers tend to prefer interventions and watershed technologies with fast returns (Chimdesa 2016). Hence, many argue that integrated watershed management, with its high initial costs, is a long-term investment that prevents small-scale, resource-poor farmers from obtaining short-term benefits (Mekonnen and Fekadu 2015). Farmers in densely populated areas with small land holdings who need communal lands to graze their herds are consequently reluctant to implement measures on lands that limit access to feed for their herds, timber and fuel wood (Mekonnen and Fekadu 2015). The land shortage requires additional investments for agricultural technologies, such as improved crops, forage, animal breeds and farming practices to compensate for the immediately lost income. Many of the successful watersheds implemented these additional investments.

Policy aspects

Over the past 20 years, the Ethiopian government has developed a policy framework to promote agriculture as a driver of economic development. The basic approach has to a large extent shifted from top-down infrastructure solutions to community-based approaches. Currently, there is a supportive policy and legal framework in the form of policies that facilitate decentralized and participatory development and institutional arrangements (IWMI 2012).

Major policies and programmes implemented in the last 20 years to facilitate integrated watershed management were:

1. The conservation-based, agricultural-development-led industrialization strategy was formulated in 1994 by the national government and is still the main policy in Ethiopia. It considers agriculture as the country's growth engine, putting smallholder farmers at the core of the strategy (Gudeta 2009). Agricultural productivity is promoted through market access, credit services and training for farmers to encourage micro- and small-scale enterprises.
2. Participatory watershed management was initiated by the government at the end of the 1990s to promote sustainable water and land resources management based on partnerships with the community (farming society) (German et al. 2007). The approach emphasizes improving the productivity of water and land resources in an ecologically and institutionally sustainable way (Gebregziabher et al. 2016).
3. The Environmental Protection, Land Administration and Use Authority is the regional equivalent of the federal Environmental Protection Authority, which was established in 1994. The major roles and responsibilities of the regional authority include ensuring the sustainable protection, development and utilization of resources, and the adherence to federal and regional policies for management, administration and use of rural land (ARD 2004). In Tigray, the regional authority is established under the Bureau of Agriculture and Rural Development, where one of its main activities is land certification, and therefore has offices in all districts.
4. The programme Managing Environmental Resources to Enable Transitions to More Sustainable Livelihoods was part of the World Food Programme's Ethiopia programme. It started in 2003 and aimed at enabling development and improving livelihoods and food security opportunities for the most vulnerable, particularly women-headed households, through the sustainable use of the natural-resources base (Tongul and Hobson 2013). The programme focusses on incorporating traditional knowledge about farming practices into the interventions.

These policy directives and programmes called for collaboration with non-governmental organizations on community-based watershed management interventions (Yaebiyo et al. 2015) and encouraged public agencies at all levels to work together (IWMI 2012). Furthermore, they promoted approaches to natural-resources management that reflected customary laws (bylaws) and tenure practices through initiating land-tenure certification procedures for farmlands, which encouraged some sense of ownership and thereby investment in natural-resources management. Although land-tenure certification procedures have been initiated for farmland, some propose alternatives to the use and management of communal lands under state ownership (Gorfu 2016). Overall, these kinds of watershed management programmes are based on lessons learned over the past several decades and have offered new opportunities to reduce farmers' dependence on rainfed, low-productivity subsistence agriculture; reverse land degradation and increase local participation.

Lessons learned from the Tigray cases

Despite the high investment costs, integrated watershed management enabled farm diversification, bringing new livelihood opportunities. The activities contributed to natural-resources management where soil and water conservation measures had significant direct food security benefits, evaluated in terms of quantity, duration and nutritional diversity. This raised local awareness and knowledge about natural-resources management and about agriculture and irrigation techniques. It also meant that women could use time more efficiently and take part in community decision-making and income-generating activities, which contributed to children's education and reduced dropout rates (UNCCD 2017). The results from the watersheds in this study show how strong community involvement and technical support can generate multiple benefits. Drawing on the lessons learned from Tigray, these models for restoration of degraded land set achievable examples for elsewhere in semi-arid Africa. For example, the combination of water harvesting methods and integrated aquaculture system can create new rural jobs and improve nutrition (Matolla Chapter 5). Realizing the full benefits at the watershed level of such investments requires external financial investment and technical support as well as social capital and collective action to complement investments at the household level. Ethiopia has shown that this can be achieved (Chisholm and Woldehanna 2012).

References

[ADF] African Development Fund. 2005. Ethiopia: Rural Water Supply and Sanitation Programme. Appraisal Report. Infrastructure Department, North, East and South, Onin. 73p. www.afdb.org/fileadmin/uploads/afdb/Documents.

[ARD] ARD Inc. 2004. Ethiopia Land Policy and Administration Assessment: Final Report with Appendices. Submitted to USAID/Ethiopia. Burlington VT. 110p. https://rmportal.net/library/content/land-tenure-and-property-rights-documents/ethiopia-land-policy-and-administration-assessment/view.

[AWS] AgWater Solutions. 2012. Watershed management in Ethiopia. Agricultural water management learning and discussion brief. http://awm-solutions.iwmi.org/Data/Sites/3/Documents/PDF/Country_Docs/Ethiopia/watershed-management-in-ethiopia.pdf.

[CCME] Canadian Council of Ministers of the Environment. 2016. Summary of Integrated Watershed Management Approaches across Canada. Canadian Council of Ministers of the Environment. 27p. ISBN: 978-1-77202-034-2.

Chisholm N, Woldehanna T. 2012. Managing watersheds for resilient livelihoods in Ethiopia. Development Co-operation Report 2012: Lessons in linking sustainability and development. OECD Publishing. Paris. https://doi.org/10.1787/dcr-2012-15-enChimdesa G. 2016. Historical Perspectives and Present Scenarios of Watershed Management in Ethiopia. *International Journal of Natural Resource Ecology and Management* 1:115–127. doi:10.11648/j.ijnrem.20160103.17.

[CSA] Central Statistical Agency of Ethiopia. 2008. Summary and statistical report of the 2007 population and housing census results. Population size by age and sex. Central Statistical Agency of Ethiopia: Addis Ababa. 113p. www.scribd.com/doc/28289334/Summary-and-Statistical-Report-of-the-2007.

Debalkew A. 2014. Management and its benefits for landless youth: the case of Helvetia's Sun Rise Project, Messebo Watershed, Tigray, Ethiopia. MSc thesis in Agroecology and Sustainable Development, Mekelle University, Mekelle, Ethiopia. 73p.

Deressa TT, Hassan RM. 2009. Economic impact of climate change on crop production in Ethiopia: evidence from cross-section measures. *Journal of African Economies* 18(4):529–554. https://doi.org/10.1093/jae/ejp002.

Descheemaeker K, Raes D, Nyssen J, Poesen J, Haile M, Deckers J. 2009. Changes in water flows and water productivity upon vegetation regeneration on degraded hill slopes in northern Ethiopia: a water balance modelling exercise. *The Rangeland Journal* 31: 237–249. doi:10.1071/RJ09010.

[EWAO] Endamehoni Woreda Agricultural Office. 2013. Business plan for the Embahsty Micro – Watershed in Endamehoni Woreda, Southern Tigray, Ethiopia. 37p. www.cmpethiopia.org.

[FAO] Food and Agriculture Organization of the United Nations. 1986. Highlands Reclamation Study, Ethiopia, Final Report, Vol. I, FAO, Rome. 354p.

Gebregziabher G, Abera DA, Gebresamuel G, Giordano M, Langan S. 2016. An assessment of integrated watershed management in Ethiopia. International Water Management Institute (IWMI). Working Paper 170, Colombo, Sri Lanka. doi:10.5337/2016.214.

Gebremeskel K. 2018. The role of integrated watershed management on improving soil health: the case of Hintalo Wejerat District, Northern Ethiopia. MSc thesis in Tropical Land Resources Management. Mekelle University, Mekelle, Ethiopia. 76p.

Gebremichael Y, Waters-Bayer A. 2007. Trees are our backbone. Integrating environment and local development in Tigray Region, Ethiopia. International Institute for Environment and Development (IIED). Issue paper no. 145. Irish Aid. Russel Press, Nottingham. http://pubs.iied.org/12539IIED/.

Gebreyohannes G, Nyssen J, Poesen J, Bauer H, Merckx R, Haile M, Deckers J. 2012. Land reclamation using reservoir sediments in Tigray, northern Ethiopia. *Soil Use and Management* 28:113–119.

German L, Mansoor H, Alemu G, Mazengia W, Amede T, Stroud A. 2007. Participatory integrated watershed management: evolution of concepts and methods in an ecoregional program of the eastern African highlands. *Agricultural Systems* 94(2):189–204.

[GIZ] Deutsche Gesellschaft für Internationale Zusammenarbeit. 2015. GIZ Ethiopia: lessons and experiences in Sustainable Land Management. GIZ support for Ethiopia's Sustainable Land Management Program, Addis Ababa, Ethiopia. 236p.

Gorfu SA. 2016. The impact of Ethiopian land certification on land conservation, maintenance and tree planting. MSc thesis. Norwegian University of Life Science. 36p. http://hdl.handle.net/11250/2403898.

Gudeta Z. 2009. How successful the Agricultural Development Led Industrialization Strategy (ADLI) will be leaving the existing land holding system intact − a major constraint for the realization of ADLI's Targets? *Ethiopian e-Journal for Research and Innovation Foresight* 1(1):19–35.

Hadush M. 2015. The role of community based watershed management for climate change adaptation in Adwa, Central Tigray Zone. *International Journal of Weather, Climate Change and Conservation Research* 1(1):11–35.

Haregeweyn N, Berhe A, Tsunekawa A, Tsubo M, Meshesha D. 2012. Integrated watershed management as an effective approach to curb land degradation: a case study of the enabered watershed in Northern Ethiopia. *Environmental Management* 50:1219–1233. doi:10.1007/s00267-012-9952-0.

Hurni H. 1993. Land degradation, famine and resource scenarios in Ethiopia. *In*: D Pimentel (ed.). *World Soil Erosion and Conservation*. Cambridge University Press, Cambridge. P 27–62.

Ilstedt U, Bargués Tobella A, Bazié HR, Bayala J, Verbeeten E, Nyberg G, Sanou J, Benegas L, Murdivarso D, Laudon H, Sheil D, Malmer A. 2016. Intermediate tree cover can maximize groundwater recharge in the seasonally dry tropics. *Nature Scientific Reports* 6:21930.

[IWMI] International Water Management Institute. 2012. Watershed management in Ethiopia. Agricultural water management learning and discussion brief. AgWater Solutions IWMI. Addis Ababa. 2p. http://awm-solutions.iwmi.org/Data/Sites/3/Documents/PDF/Country_Docs/Ethiopia/watershed-management-in-ethiopia.pdf.

Kirubel M, Gebreyesus B. 2011. Impact assessment of soil and water conservation measures at Medego watershed in Tigray, Northern Ethiopia. *Maejo International Journal of Science and Technology* 5(3):312–330.

Mekonnen GT, Fekadu A. 2015. Experiences and challenges of integrated watershed management in Central Zones of Southern Ethiopia. *International Journal of Current Research* 7:20973–20979.

Negusse T, Yazew E, Tadesse N. 2013. Quantification of the impact of integrated soil and water conservation measures on groundwater availability in Mendae Catchment, Abraha We-Atsebaha, eastern Tigray, Ethiopia. *Momona Ethiopian Journal of Science* 5:117–136.

Nyberg G, Knutsson P, Ostwald M, Oborn I, Wredle E, Otieno D, Mureithi S, Mwangi P, Said MY, Jirström M, Grönvall A, Wernersson J, Svanlund S, Saxer L, Geutjes L, Karmebäck V, Wairore JN, Wambui R, De Leeuw J, Malmer A. 2015. Enclosures in West Pokot, Kenya: transforming land, livestock and livelihoods in drylands. *Pastoralism: Research, Policy and Practice* 5:25. doi:10.1186/s13570-015-0044-7.

Nyssen J, Naudts J, De Geyndt K, Mitiku H, Poesen J, Moeyersons J, Deckers J. 2008. Soils and land use in the Tigray Highlands (Northern Ethiopia). *Land Degradation & Development* 19:250–274. doi:10.1002/ldr.840.

Nyssen J, Poesen J, Haile M, Moeyersons J, Deckers J, Hurini H. 2009. Effects of land use and land cover on sheet and rill erosion rates in the Tigray high lands, Ethiopia. *Zeitschrift fur Geomorphologie* 53:171–197. doi:10.1127/0372-8854/2009/0053-0171.

Smur G. 2017. The impact of small scale irrigation on women headed households' economic and social empowerment: the case of Kilte-Awlaelo District, Tigray, Ethiopia. MA thesis in Gender and Development Studies. Mekelle University, Ethiopia. 83p.

Teka K, Haftu M. 2012. Land suitability characterization for crop and fruit production in Midlands of Tigray, Ethiopia. *Momona Ethiopian Journal of Science* 4(1):64–76.

Teka K, Haftu M, Ostwald M, Cederberg C. (under review). Role of integrated watershed management in reducing soil erosion and livelihood improvement: the case of Gulle Watershed, Northern Ethiopia. *Journal of Land Degradation and Development*.

Teka K, Nyssen J, Teha N, Haile M, Deckers J. 2015. Soil, land use and landform relationship in the Precambrian lowlands of northern Ethiopia. *Catena* 131:84–91.

Teka K, Van Rompaey A, Poesen J. 2013. Assessing the role of policies on land use change and agricultural development since 1960s in Northern Ethiopia. *Land Use Policy* 30:944–951.

Teka K, Van Rompaey A, Poesen J, Van Bruyssel S, Deckers J, Amare K. 2014. Spatial analysis of land cover changes in Eastern Tigray (Ethiopia) from 1965 till 2007: are there signs of a forest transition? *Land Degradation & Development* 26:680–689. doi:10.1002/ldr.2275.

Tesfaye A, Brouwer R, van der Zaag P, Negatu W. 2016. Assessing the costs and benefits of improved land management practices in three watershed areas in Ethiopia. *International Soil and Water Conservation Research* 4:20–29. doi:10.1016/j.iswcr.2016.01.003.

Tongul H, Hobson M. 2013. Scaling up an integrated watershed management approach through social protection programmes in Ethiopia: the MERET and PSNP schemes. Hunger – Nutrition – Climate Justice a New Dialogue: Putting People at the Heart of Global Development. Case studies: policy responses, local to national. 4p. Mary Robinson Foundation Climate Justice. Dublin. www.mrfcj.org/wp-content/uploads/2015/09/2013-04-16-Ethiopia-MERET.pdf.

[UNCCD] United Nations Convention to Combat Desertification. 2017. *Ethiopia's Tigray Region Bags Gold Award for Greening*. Published 26 August 2017. African Independent. Cape Town. www.africanindy.com/environment/ethiopias-tigray-region-bags-gold-award-for-greening-10919650.

[WFP] World Food Program. 2012. The Role of Food Assistance in Social Protection. Update of WFP's Safety Nets Policy. World Food Program. Rome Italy. 36p. https://documents.wfp.org/stellent/groups/public/documents/communications/wfp254438.pdf.

Woldearegay K. 2015. Water Harvesting and Climate Change Adaptation: the experience of Tigray region in Ethiopia. Training and experience sharing of the AR Project; 20 April 2015, International Livestock Research Institute (ILRI), Addis Ababa, Ethiopia. 64p. www.rainfounadtion.org.

Yaebiyo G, Tesfay Y, Assefa D. 2015. Socio-economic impact assessment of integrated watershed management in Sheka Watershed, Ethiopia. *Journal of Economics and Sustainable Development* 6:202–212. ISSN 2222-1700.

Zenebe A, Vanmaercke M, Poesen J, Verstraeten G, Haregeweyn N, Haile M, Amare K, Deckers J, Nyssen J. 2013. Spatial and temporal variability of river flows in the degraded semi-arid tropical mountain of Northern Ethiopia. *Zeitschrift für Geomorphologie* 57:143–169. doi:10.1127/0372-8854/2012/0080.

7 Smallholder maize-based systems
A piece of the puzzle for sustaining food security in Nigeria

Julius B Adewopo

Status of maize-based systems

Maize is the backbone of food security across Sub-Saharan Africa (Shiferaw *et al.* 2011), accounting for up to half of the calorie intake (Nuss and Tanumihardjo 2011), and a core ingredient in animal feed. Almost all the continent's total maize output (96 per cent) comes from 20 countries, with Nigeria at the top with 15 per cent of African maize production or 10.4 million tonnes (FAOSTAT 2018). The importance of staple crops for food security can be viewed against Nigeria's population trends, which went from 45 to over 190 million between 1960 and 2017 (WB 2017), among the fastest growth in the world and projected to double by 2050 (IF 2019).

In Sub-Saharan Africa, about 70 per cent of maize cultivation is done by smallholder farmers (Macauley 2015; Smale *et al.* 2011) who depend on it for both their subsistence and livelihoods. Many smallholder farmers' maize yields are one-tenth those of average yields for the United States (Figure 7.1). The latter can leverage economies of scale that are unavailable to African smallholders, whose holdings range between 0.2 and three hectares and are often spread across small scattered parcels. Instead, maize-based multiple-cropping systems have evolved as livelihood strategies in response to remoteness, where poor transport and infrastructure hinder marketing opportunities and access to extension services. Here, local small-scale markets have developed, where maize and similar staple crops can be locally processed and stored.

Maize in Nigeria and Africa at large

Maize was introduced to Nigeria in the fifteenth century (Blench 1997). It was cultivated as a subsistence crop and gradually evolved into a commercial crop providing raw materials for agro-industries, such as grains for animal feeds, processed cereal, and beer (Ammani 2015; Iken and Amusa 2004).

Maize production first started in the humid forest zones in the south. While cassava (Onoja Chapter 4) remains the main crop in the southern

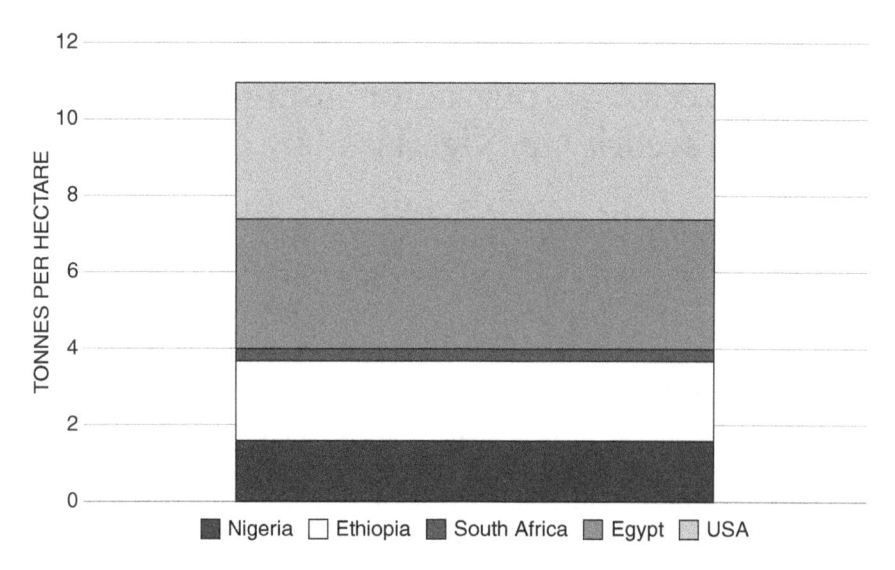

Figure 7.1 Average maize yields in 2016 in selected countries.
Source: FAOSTAT 2018.

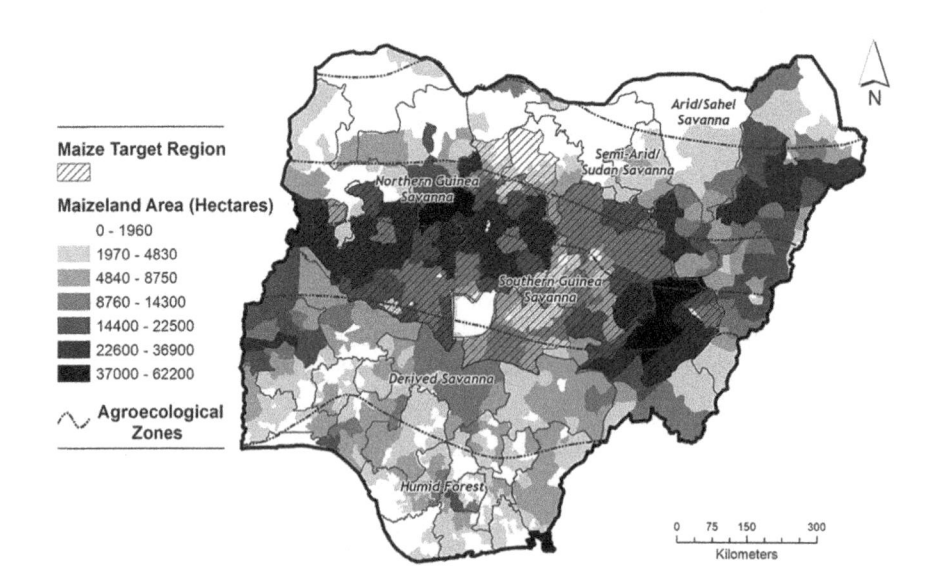

Figure 7.2 Major maize-producing regions in Nigeria.
Sources: Author's adaptation from HarvestChoice 2015a, b.

to central forest zones, maize has shifted northwards into the Guinea and Sudan savanna agroecological zones across the middle of the country since the 1970s (Blench 1997; Figure 7.2). Maize is suitable across diverse altitudes and latitudes, however, compared to the humid zones, the savanna has more favourable growing conditions, particularly less cloud cover (more solar radiation), suitable rainfall ranging from 700 to 1050 millimetres annually, and a terrain that enables livestock production to be combined with field operations (Kim et al. 1993; Obi 1991).

Although among the top producers in Africa (FAOSTAT 2018), Nigeria has marginally been a net maize importer. According to the 2016–2020 policy and strategy document for agricultural promotion, maize is the only one of the 13 listed food crops and products where supply closely matches national demand (93 per cent; FMARD 2016). Historically, Nigeria's food production deficit and inability to meet the increasing domestic grain demand were linked to inadequate input supply and poor extension support (Liverpool-Tasie et al. 2017). Food security is not only challenged by market failures, lack of support and a growing population but also by insurgents and conflicts. In the spring of 2019, the food insecurity situation was deemed 'stressed' in at least ten states, and at 'crisis' and 'emergency' phases in the north-eastern states bordering Cameroon, Chad and Niger, using the Integrated Food Security Phase Classification (FEWSNET 2019).

First, Nigeria's maize production is summarized in a historical policy development context over five periods, to frame the context of maize-based multiple-cropping systems.

1970s – Multiple-cropping systems

Figure 7.3 shows stagnating trends for maize production and area in the 1960s and declining trends in the 1970s. In 1972, the National Accelerated Food Production Programme was launched and in 1976 Operation Feed the Nation. These two policies can be described as revolutionary, but the impacts remain debatable. They served as precursors of subsequent policies that resulted in an expansion of the area used for maize. Broadly, these policies favoured maize in multiple-cropping systems by providing input subsidies on major grain and legume crops (mainly maize, cowpea, and soybeans) and by encouraging the establishment of farms and gardens on any available nearby land. Furthermore, the River Basin and Rural Development Authorities were established in 1976 with the mission to accelerate rural development through support for year-round production under irrigated and rainfed systems. This provided an advantage to farmers in the savanna region, where most of these basins were established, by allowing for year-round maize production intercropped with other seasonal crops, including vegetables, spices and legumes.

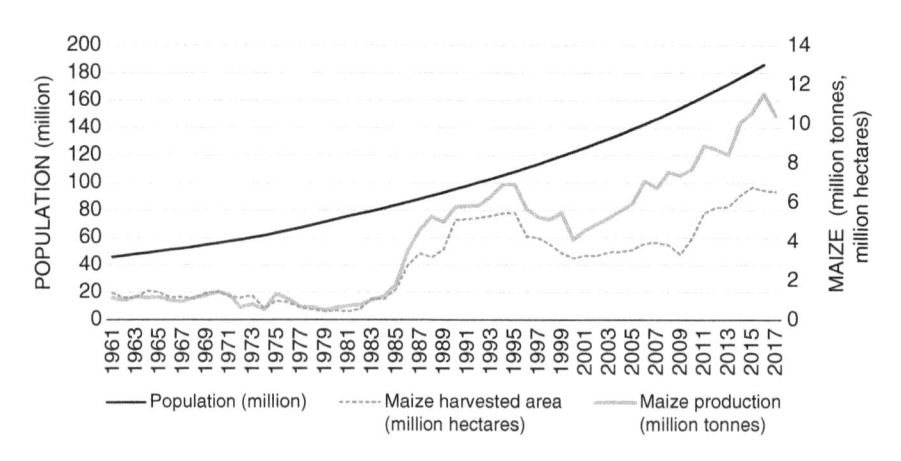

Figure 7.3 Population, maize area and maize production in Nigeria 1961–2017.
Source: FAOSTAT 2018.

1980–1992 – *First expansion*

The first major expansion of maize production in Nigeria coincided with the Green Revolution in the early 1980s. Policies targeted improved access to inputs through subsidies and credits and aimed to reverse the declining trend of national agricultural productivity in cereals and pulses (Adeyemo 1984; Hassan *et al.* 2014). Between 1987 and 1992, the annual total maize production increased from 0.4 to 5.7 million tonnes and resulted in a simultaneous drop in maize imports from 347 to 0.3 million tonnes (FAOSTAT 2018). However, this production increase was largely due to an unprecedented increase in the total maize area, by converting 4.5 million hectares of previously uncultivated land (Figure 7.3).

1992–2002 – *Stagnation*

In 1992 the Nigerian Agricultural Land Development Authority was established with the mission to provide support for agricultural expansion through provision of funds to agricultural programmes, facilitation of input sourcing and procurement, acquisition of machinery and training of agricultural programme staff. Through the Authority, farmers were organized into cooperative societies and farmer groups for ease of access to credit and training, with the expectation that this would translate into improved support for rural farmers, especially those in proximity to previously established infrastructure such as the River Basins. Similarly, the National Fadama Development Project and the World Bank-funded Agricultural Development Project were initiated in the early 1990s. Bureaucracy and poor technical oversight have meant that these interventions are rife with shortcomings (Akinsola and

Oladele 2004; Uche 2011), with minimal gains in maize productivity or cultivated maize area during the period.

2002–2007 – Maize boom

In 2002 the National Special Programme on Food Security was launched and focussed more on providing general support to encourage farming than on promoting maize production. Nevertheless, this seems to have benefitted maize, and production increased from 4.0 to 7.6 million tonnes within five years (Figure 7.3). Maize gained popularity because it was compatible with many environmental conditions as well as other crops and because it offered a fast return-on-investment, which met the needs of households. This policy effect continued after the policy ended in 2008, reaching 2.1 tonnes per hectare in 2009 (Cadoni and Angelucci 2013).

2009 – ongoing – Second expansion

The Agricultural Transformation Agenda programme launched in 2009, introducing new fertilizer support with a focus on improving farmers' access to quality fertilizer at lower cost, especially during the main cropping season (see also Onoja Chapter 4). Between 2009 and 2014, the harvested maize area increased from 3.4 to 5.9 million hectares, which increased production from 3.3 to 6.8 million tonnes (Figure 7.3; FAOSTAT 2018). Furthermore, although farms larger than ten hectares do exist, up to 80 per cent of the Nigerian maize remains predominantly cultivated in multiple-cropping systems on small fragmented plots (Onuk *et al.* 2015). In 2016, the Agricultural Transformation Agenda was modified to become the Agriculture Promotion Policy (FMARD 2016). This policy attempts to redirect government efforts to address some major deficiencies of previous programmes, including engagement of stakeholders, leveraging digital technologies and prioritizing poverty reduction among farmers.

Comparisons of maize development in Africa

Productivity remains a challenge for Nigeria, as it is for the neighbouring countries. Nationally, maize yields are around two tonnes per hectare, while the potential yield is more than four times that, about 8.6 tonnes per hectare (Olaniyan and Lucas 2004). Nigeria's average maize yield is half of the yields in South Africa and Ethiopia, and one-fifth of that in Egypt (Figure 7.1). There are several explanations for the yield gaps. First, like in many Sub-Saharan African countries, most Nigerian maize is rainfed. Second, comparatively less land was required to achieve the production increase in Egypt and Ethiopia, which suggests that as land was available, the need to develop land-efficient technologies was less of a driver in Nigeria (Figures 7.1, 7.4a, b). Ethiopian maize remains rainfed; however

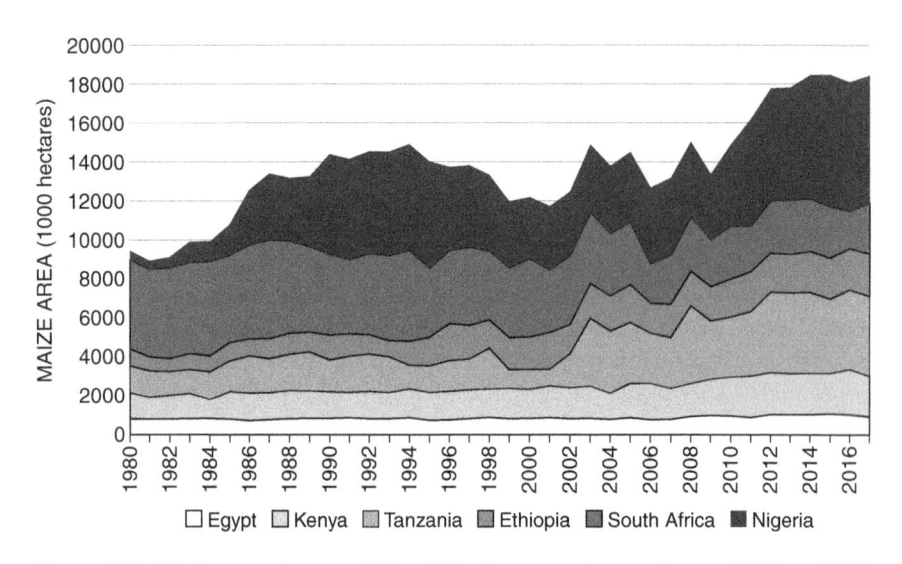

Figure 7.4a Maize area harvested for Africa's top maize producers 1980 to 2017. Unit: 1000 hectares.

Source: FAOSTAT 2018.

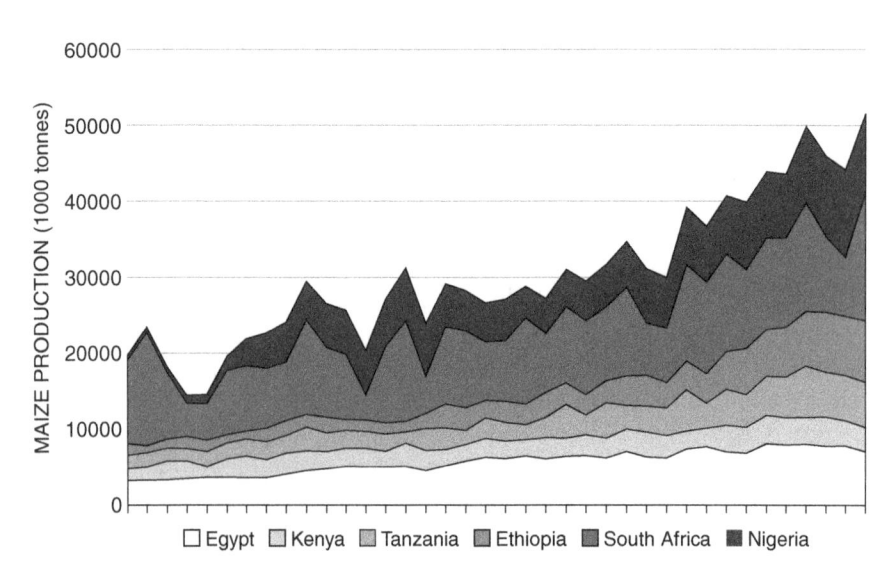

Figure 7.4b Maize production for Africa's top maize producers 1980 to 2017. Unit: 1,000 tonnes. See also Nigeria in Figure 7.3.

Source: FAOSTAT 2018.

after the famine in the mid-1980s, the government has invested in research, development and extension to find suitable higher-yielding varieties, nitrogen and phosphorous fertilizers, and in converting some teff and sorghum areas to maize (Abate et al. 2015). South Africa has similar average yields to Ethiopia but produces rainfed high-yielding varieties of white maize with large interannual variability due to droughts (FAO 2018). Moreover, in Ethiopia domestic demand is fuelled by the population increase, while as South Africa also supplies large parts of Southern Africa, the unstable production has a large impact on regional food prices and food security. In contrast, the maize area in Egypt has been relatively stable at about one million hectares (Figure 7.4a) for the past 50 years, while yield increases are predominantly attributed to intensified use of surface irrigation, high-yielding varieties, and fertilizer (Zohry et al. 2017). For example, according to FAOSTAT 2018, in Egypt the average fertilizer use for the period 2002–2015 was 594 kilogrammes per hectare, compared to eight kilogrammes per hectare in Nigeria. While the accuracy of these numbers may be debatable, the magnitude of the difference is instructive.

Maize in multiple-cropping systems

The land area of Nigeria is 91 million hectares, of which 39 per cent is classified as arable land, while permanent crops and forest resources make up 7 and 9 per cent, respectively. In 2016, maize was cultivated on 6.5 million hectares (FAOSTAT 2018), which may be a conservative estimate since it is unclear how maize intercropped with other crops, such as cowpea and groundnut, is accounted for in the national statistics. In the 1970s and 1980s, researchers estimated that 99 per cent of cowpea, 95 per cent of groundnut, 90 per cent of sorghum and millet, and 75 per cent of maize grown in Nigeria was intercropped (Ofori and Stern 1987). Recent statistics on intercropping practices are unavailable. With the emergence of large-scale producers, who primarily practice monoculture, multiple-cropping systems' share of the total maize production has likely decreased. Some estimates from the Nigerian savanna region state that one out of every five farmers now practices maize as monoculture, while the rest combine maize with other crops (Mustapha and Salihu 2015). Furthermore, an unpublished agronomic pilot survey of 780 farmers in Kano, Kaduna, and Katsina states in the Guinea and Sudan savanna (IITA 2016) indicated that three out of every five farmers intercropped maize as a general practice, and about four out of five mixed maize with other crops during the last three years of the maize rotation (Figure 7.5).

Figure 7.5 A typical mixed maize-based system with okra, soybeans and cowpea on *c.*1 to 2 hectares of farmland in Doguwa, Kano State, in the Sudan Savanna agroecological zone, Nigeria.

Photo credit: Adewopo 2017.

Factors for success in studied cases

Table 7.1 highlights some of the reviewed research conducted on various maize-based systems in Nigeria since the 1970s as examples of success. These examples demonstrate a focus on productivity and inputs towards a more efficient use of resources with interaction effects.

Diversification of products

An important explanation for the popularity and fast expansion of maize-based systems is that maize can easily be planted within existing farming systems and that it offers higher yields than traditional grain crops (Macauley 2015). Although the savanna region supplies 65 per cent of Nigeria's maize production, it remains a low-input system with widespread intercropping practices. For instance, in the northern region maize is mixed with legumes such as soybean, cowpea or groundnut, or cereals, such as sorghum and rice, while in the southern region, maize is intercropped with cassava (Onoja Chapter 4) and yam (Thayamini and Brintha 2010). As maize matures in succession, it is suitable for intercropping with tuber crops such as sweet potato, and vegetables such as tomato, onion and pepper.

Table 7.1 Indicators studied in maize-based multiple-cropping systems in Nigeria

Main farming systems studied	Main benefits studied	Reference
Maize intercropped with groundnuts, sorghum, and millet, Northern Nigeria	Profitability (increased), as indicated by cash return	Baker 1978
Maize in alternated intercropping with millet and sorghum, Northern Nigeria	Yield (increase) of maize in alternate intercropping compared to monoculture	Baker 1979
Maize intercropped with cowpea sequential cropping on intensively cultivated tropical Ultisol, Abeokuta, Nigeria	Yield and net profitability, improved nitrogen uptake of maize on poor soils	Adetunji 1996
Maize after soybean, Guinea savanna, Nigeria	Micro-nutrient uptake of maize after legume: maize yield	Carsky *et al.* 1997
New intensive system with maize in rotation with soybeans and livestock, dry savanna, northern Nigeria	Resource management of new germplasm; income, production, and land area	Sanginga *et al.* 2003*b*
Maize cultivated with cowpea, groundnut, or soybean; soybean with cowpea or groundnut, Zaria, Nigeria	Land-use efficiency based on farm size and production	Herbert 2005
Maize intercropped with cowpea, south-western Nigeria	Input optimisation, biological nutrient fertiliser effects of cowpea	Amujoyegbe *et al.* 2008
Maize intercropped with cowpea, north-central region, Nigeria	Technical efficiency of maize intercropped with cowpea, based on gross margin	Onuk *et al.* 2015

Soil nutrient management

According to one study, maize production under the current situation is optimal on 28 per cent of African agricultural land, suitable on 59 per cent, and unsuitable on 13 per cent (Peter *et al.* 2017). The same study concluded that intercropping to utilize biological nitrogen fixation can benefit areas that are suboptimal for maize. Biological nitrogen fixation is a process in which organisms in symbiosis with certain plants, such as legumes, convert atmospheric nitrogen to ammonia, which crops can assimilate (Wagner 2011). This can improve soil fertility and reduce nitrogen fertilizer requirements for subsequent non-legume crops (Table 7.1, see for example Sanginga *et al.* 2001, 2003*a*). For example, one field trial showed increases in maize yields by 16 to 32 per cent, when planted directly after soybeans (Carsky *et al.* 1997).

Diversified incomes

As high-yielding maize varieties require a higher supply of nitrogen than local varieties (Onwueme and Sinha 1991), the degraded and nutrient-poor soil conditions prevalent in most Nigerian croplands limit the potential to optimize yields (Adetunji 1996; Giller *et al.* 2011). Despite the past efforts to develop drought-tolerant higher-yielding varieties and fertilizers for nutrient-poor soils (Binswanger-Mkhize and Savastano 2017; Liverpool-Tasie *et al.* 2017) for monoculture systems, smallholder farmers seem to prefer to cultivate maize in traditional ways with other crops (see example in Figure 7.5). By managing rotation and intercropping on multiple plots within one farm holding, farmers can often optimize the allocation of resources (labour and capital) within the season and improve farm-level technical efficiency (Adetunji 1996; Amujoyegbe *et al.* 2008; Awotide and Agboola 2014; Sanginga *et al.* 2003b). Smallholder farmers with less than five hectares are relatively flexible and can make intra-seasonal changes. Some studies suggest that farmers were able to optimize return-on-investment on smaller farmlands by adopting maize-based multiple-cropping (Sanginga *et al.* 2003a,b) and intercropping practices to maximize returns and economic flexibility, under prevalent circumstances of poor access to infrastructure and financial resources, and uncertain land tenure and user rights (Makinde *et al.* 2011; Quainoo *et al.* 2000). Making available shorter-term varieties can help farmers take more flexible and adapted decisions as seasonal climatic situations vary. Also, the yield and net profitability can be strategically improved on nutrient-impoverished soils through compensatory nutrient dynamics of the constituent crops (Adetunji 1996; Onuk *et al.* 2015).

Social, economic and environmental co-benefits

Each year, crop residues from millions of hectares provide additional benefits such as soil quality amelioration, construction materials for low-cost thatch roofs, fodder for livestock and fuel for cooking, especially in savanna areas where trees are sparsely distributed (Olaniyan 2015). These benefits often incentivize farmers to continue to cultivate maize lands for household consumption and contribute to national maize grain production. Scientists in Egypt have studied environmental functions, such as different root depths and root biomass, and experimented with maize in rotations and intercropping systems with legumes, forage and fruit trees to identify new systems that benefit yields of all crops, reduce land, water and fertilizer use, and control weed and pests (Zohry and Ouda 2017; Zohry *et al.* 2017).

In summary, the major advantages of maize-based multiple-cropping systems accrue to farm-level resource use-efficiency, such as improved nutrient management, reduced labour input per unit area, and reduced transportation cost per unit produced (Tables 7.1 and 7.2).

Limitations

Scale

Land fragmentation can be discussed both as a cause and an effect of small-scale multiple-cropping systems. The potential negative impacts of smallholder maize-based systems are primarily linked to economies of scale (Table 7.2). Certain multiple-cropping systems may limit the use of modern technologies on small-scale farms. For example, intercropping or dense and multi-level canopy structures can be incompatible with machinery for

Table 7.2 Benefits and potential drawbacks of smallholder maize-based multiple-cropping systems in Nigeria

	Benefit	*Potential drawbacks*
Soil	Nutrient management efficiency through legume-induced biological nitrogen fixation and improved nutrient cycling though farm residue incorporation Improved soil quality (microclimate, tilth, organic matter, structure)	Inefficient nutrient management may encourage maintained status quo in production or overlooked yield decline Indirectly incentivises land fragmentation
Economics	Increased return on investment by harnessing multiple crop yields Low investment cost to establish and generate food and livelihoods Diversification as risk reduction strategy of harvest time	Difficult to apply farm technology and economies of scale Labour intensive, with potential implications for women and children labourers in some cultures
Food security	Diversified household nutritional intake and diets (legumes, vegetables, spices) Provision of other materials, e.g. feed, fuel, and shelter materials	May not be compatible with yield optimisation or yield-gap minimisation for main crops, including maize
Crop and technical knowhow	Crop diversity reduces the risk for pest- and disease-related losses Improved weed control after crop establishment	System-level knowledge of crop interactions and optimal thresholds of management practices are critical to balance risks and rewards Requires well-developed extension support

basic operations such as weeding, thinning and harvesting. Such systems are therefore often perceived to disfavour practices aiming at economic efficiency, such as mechanization and land mergers. Small-scale farms also rely on household members for permanent and temporary farm work, which often has implications for children's opportunities to attend school and women's participation in meetings and networks, which can have more impacts on farm productivity and household incomes in the long run.

The prevalent practice of dividing farmlands and allotting plots to entitled family members as inheritance reinforces fragmentation and subsistence (McPherson 1983; Simmon 1987). A major downside of smaller plots is farmers' reluctance to test new agriculture solutions that require standard spacing between crops or trees. As smallholder farmers are often poorly equipped to manage soil nutrient balance in their farmlands (Giller *et al.* 2011), they may seek more fertile lands in forests and protected land. Fragmentation may therefore also result in encroachment into forestlands and protected areas. Although multiple-cropping systems have supported subsistence and household food security in the past, achieving ecologically sustainable systems will require new solutions to halt fragmentation and either aggregate farming practices across plots, or aggregate farmlands (Iheke and Amaechi 2015; Okezie *et al.* 2012).

Furthermore, technology adoption is constrained by gender imbalances in terms of land tenure (for example Pretty 2008) and limited access to cash and credit (AfDB 2015). As in many rural areas in Africa, low literacy levels and inefficient extension systems limit outreach on agricultural information. Moreover, many farmers are women who have less direct access to land, resources, and information updates than men, which often leads to misinformed management decisions. Some studies found that women-headed households had lower yields than those headed by men (a pattern also found in Onoja Chapter 4), likely associated with poorer households needing to work extra for wealthier farmers for cash, usually right when they need to work the most on their own farms (Peter *et al.* 2017).

Inconsistent input support

Although past government policies aimed at improving access to seeds and fertilizers, inconsistent fertilizer policies have often favoured either monopoly or liberalization at different times (Nagy and Edun 2002; Oko 2011). For instance, the discontinuation of the national fertilizer subsidy and distribution programme between 1997 and 2002 led to a 50 per cent increase in fertilizer prices, with a consequent sharp decline in fertilizer use (Oko 2011); the area cultivated with maize declined from 5.2 to 4.2 million hectares (Cadoni and Angelucci 2013). Despite this, maize production remained relatively unchanged and consequently national maize yields increased slightly from 1.2 to 1.5 tonnes per hectare between 1997 and

2002 (FAOSTAT 2018; Figure 7.3). Little research-based documentation exists to fully elucidate coping strategies that have been adopted by smallholder farmers during periods when optimal fertilizer usage is cost-prohibitive due to shifting government policies. Some scientists argue that the deregulation of input markets and provision of fertilizer credits to farmers, starting in the 1980s, unintentionally resulted in increased cultivation of natural lands rather than in the expected intensified production on existing croplands (Binswanger-Mkhize and Savastano 2017; IITA 1991; Nagy and Edun 2002). This can partly be explained by the abundant availability of cheap labour and that fertilizer credits led landless people who did not have their own farm holdings to venture into previously uncultivated lands for farming, including fringes of forest reserves, buffer zones, national parks and important ecological corridors with major environmental implications.

Without incentives that nudge farmers to adopt sustainable practices, and structured policies to guide local planners and decision-makers, the fragmentation and expansion of farmlands onto previously uncultivated lands and clearing of important ecological corridors will likely continue. This raises concern about the agroecological sustainability within the savannas in general.

Policy aspects

Nationally, the most widely practised maize-based systems are those with legumes or cereals. Cropping systems with legumes were promoted in the 1970s. However, since the 1980s, no national agricultural policy or programme has directly promoted multiple-cropping systems with maize in Nigeria. Similarly, intercrops of roots and tubers have evolved in southern Nigeria, largely without policy support (Onoja Chapter 4). The co-benefits of intercropping, such as additional crops in maize-based systems, are rarely recognized in national agricultural planning or performance assessments, hence no data are reported on the presence of multiple-cropping systems in the statistics.

Since 2016, the Agriculture Promotion Policy (FMARD 2016) has been guided to deliver on three themes: productivity enhancements, private sector investment and institutional realignment. Soil fertility is considered a key element that must be addressed to achieve enhanced productivity. Therefore, the policy includes mandates for soil fertility management to improve environmental values as well as food security, for example, formal fallow periods, erosion control measures, tree planting, improved conservation, reforestation and, green belt policies. Moreover, the policy targets fertilizer quality control, the use of organic fertilizer, and an aligned strategy on fertilizer supply and demand in regions that require the most support. The strategies could include crop rotations and intercrops with suitable annual or perennial legumes where biological nitrogen fixation has positive benefits on subsequent crops, which would benefit poor farmers in particular (Peter *et al.* 2017). Interventions towards sustainable production can provide a new entry point for raising the visibility of the multiple benefits that smallholder farmers with multifunctional land uses bring to the national agenda and rural livelihoods through improved productivity.

One policy intervention that remains critical for more effective land-use pertains to land tenure and ownership. The current tenure system acknowledges communal ownership, inheritance, individual ownership, leasehold, rent, gift, free hold and tenant at the government's will, where communal ownership and inheritance are the most common. Although individual or community access to land for cultivation is allowed, the policies are often not aligned with the national land use act (Nwocha 2016), which gives government the sovereign ownership or control of land. This contributes to land fragmentation, as increasing numbers of community or family members lay claim on communally or family-held land (compare with Shomkegh Chapter 2 and Onoja Chapter 4).

There is public investment in farmers' access to training and materials that could improve farm management practices, such as seed varietal selection, fertilizer application, spacing, and timing of tending operations (Adama *et al.* 2016; Degrande *et al.* 2015). Moreover, government policies on input access typically focus on improving yields, but the net increase in production (from maize or companion crops) does not always translate into market access and higher net returns for farmers (Binswanger-Mkhize and Savastano 2017; Liverpool-Tasie *et al.* 2017). Therefore, policies should also consider strengthening farmers' access to markets and provide incentives for value-added processing of farm outputs, for instance by credit and extension support.

Lessons learned from the case

In contrast to monocultures, the contribution of Nigeria's smallholder maize-based multiple-cropping systems to national food security is insufficiently assessed and likely underestimated. For example, these farming systems are overlooked in policy initiatives, such as the 'Zero Hunger Initiative' which was implemented in 2017 and is envisioned to empower youth and rural population to produce adequate food and improved nutrition.

Agricultural policies and actions of the federal government should be guided by a clear understanding of the comparative advantage of multiple-cropping systems with maize. For instance, the strategic Anchor Borrowers Programme (CBN 2016) and Growth Enhancement Scheme (Ejiogu 2017) initiatives are implemented to encourage agricultural production and can offer incentives such as extension support and higher credit lines to farmers who adopt production practices with environmental, social, and economic benefits. So far, the Central Bank of Nigeria's Quarterly Report (CBN 2017) showed that over US$4.1 million had been disbursed to 10,260 farmers under the Agricultural Credit Guarantee Schemes, with at least 30 per cent of the credit recipients being maize (monoculture) producers. Higher production impacts could be achieved if farmers were supported in adopting practices that generate multifunctional benefits from their maize farmlands.

Past policy interventions to improve maize production in Nigeria have prioritized optimizing maize grain yields and improving tolerance or resistance to biotic and abiotic stress (Olaniyan 2015). Successful policies must be intentionally geared towards providing a blend of critical inputs, mainly fertilizer and seed, and investing in extension services for appropriate farm-level solutions (Liverpool-Tasie *et al.* 2017). National food security policies should be guided by scientific evidence on the unique characteristics of, and potentials for, crop rotations and multiple-cropping farming systems that support and strengthen small-scale farmers' contributions to multiple Sustainable Development Goals. Such evidence needs to be based on real indicators of multifunctional land uses (see examples in Tables 7.1 and 7.2), such as net profitably, land-use equivalent ratio, biological nitrogen fixation, and co-benefits.

References

Abate T, Shiferaw B, Menkir A, Wegary D, Kebede Y, Tesfaye K, Kassie M, Bogale G, Tadesse B, Keno T. 2015. Factors that transformed maize productivity in Ethiopia. *Food Security* 7(5):965–981. https://link.springer.com/article/10.1007/s12571-015-0488-z.

Adama JI, Ohwofasa BO, Ogunjobi JO. 2016. Transformation of agricultural education in Nigeria: implication for food security. *Journal of Economics and Sustainable Development* 7(7):1–8.

Adetunji MT. 1996. Nitrogen utilization by maize in a maize-cowpea sequential cropping of an intensively cultivated tropical Ultisol. *Journal of India Society of Soil Science* 44(1):85–88.

Adeyemo R. 1984. The food marketing system: implications of the green revolution programme in Nigeria. *Agricultural Systems* 14(3):143–157.

[AfDB] African Development Bank. 2015. Abuja Declaration on fertilizer for African Green Revolution. Report. Abidjan Cote d'Ivoire. www.afdb.org/en/topics-and-sectors/initiatives-partnerships/african-fertilizer-financing-mechanism/abuja-declaration/.

Akinsola OD, Oladele OI. 2004. National agricultural land development authority's (NALDA) intervention programme in Nigerian agriculture: lessons and challenges. *Food, Agriculture, & Environment* 2(1):249–254.

Ammani A. 2015. Trend Analysis of Maize Production and Productivity in Nigeria. *Journal of Basic and Applied Research International* 2(3):95–103.

Amujoyegbe BJ, Bamire AS, Elemo KO. 2008. Agronomic analysis of fertilizer effects on maize/cowpea intercropping in Ile-Ife and Abeokuta, Southwestern Nigeria. *ASSET Series* A 8(1):62–72. http://citeseerx.ist.psu.edu/viewdoc/download?doi=10.1.1.847.4986&rep=rep1&type=pdf.

Awotide DO, Agboola PO. 2014. Relationship between land fragmentation and maize farmers' productivity in northern Nigeria. *Journal of Life and Physical Sciences* 3(2):1–9.

Baker EF. 1978. Mixed cropping in Northern Nigeria I. Cereals and groundnuts. *Experimental Agriculture* 14(4):293–298.

Baker EF. 1979. Mixed cropping in Northern Nigeria. III. Mixtures of cereals. *Experimental Agriculture* 15(1):41–48.

Binswanger-Mkhize HP, Savastano S. 2017. Agricultural intensification: the status in six African countries. *Food Policy* 67:26–40.

Blench R. 1997. The history of agriculture in Northeastern Nigeria. *In*: Barreteau D, Dognin R, von Graffenried C (eds). L'homme et le milieu végétal dans le bassin du Lac Tchad. *Editions de l'ORSTOM*, Paris. P 69–112. http://horizon.documentation.ird.fr/exl-doc/pleins_textes/pleins_textes_6/colloques2/010012404.pdf.

Cadoni P, Angelucci F. 2013. Analysis of incentives and disincentives for maize in Nigeria. Technical notes series, Monitoring and Analysing Food and Agricultural Policies (MAFAP), FAO, Rome. 41p. www.fao.org/3/a-at580e.pdf.

Carsky RJ, Abaidoo R, Dashiell KE, Sanginga N. 1997. Effect of soybean on subsequent maize grain yield in Guinea savanna of West Africa. *African Crop Science Journal* 5:31–39.

[CBN] Central Bank of Nigeria. 2016. Anchor Borrowers Programme Guidelines. Development Finance Unit Central Bank of Nigeria, Abuja. 19p. www.cbn.gov.ng/out/2017/dfd/anchor%20borrowers%20programme%20guidelines%20-dec%20%202016.pdf.

[CBN] Central Bank of Nigeria. 2017. Economic Report Fourth Quarter. Abuja. 54p. www.cbn.gov.ng/Out/2018/RSD/CBN%20ECONOMIC%20REPORT%20FOURTH%20QUARTER%20%20%20%202017%20Published.pdf.

Degrande A, Tchoundjeu Z, Kwidja R, Fouepe GF. 2015. Rural Resource Centres: A Community Approach to Agricultural Extension. Note 10. GFRAS Good Practice Notes for Extension and Advisory Services. Lindau, Switzerland. www.worldagroforestry.org/downloads/Publications/PDFS/BR15624.pdf.

Ejiogu AO. 2017. Growth Enhancement Scheme (GES) of the Nigerian Agricultural Transformation Agenda: looking back and thinking ahead. *Agricultural Policy Research Network* 3(1):28–41.

[FAO] Food and Agricultural Organization of the United Nations. 2018. GIEWS Global Information and Early Warning System. Country Briefs. South Africa. Reference date: 8 November 2018. Rome. www.fao.org/giews/countrybrief/country.jsp?code=ZAF.

FAOSTAT 2018. FAO Statistical Database (online). Food and Agricultural Organization of the United Nations. Rome. http://faostat.fao.org.

[FEWSNET] Famine Early Warning Systems Network. 2019. Key message update. Over 2.0 million IDPs are in the Northeast as the Conflict Persists. West Africa, Nigeria. January 2019. http://fews.net/west-africa/nigeria/key-message-update/january-2019.

[FMARD] Federal Ministry of Agriculture and Rural Development. 2016. The Agriculture Promotion Policy (2016–2020): building on the successes of the ATA, closing key gaps. Policy and Strategy Document. Federal Ministry of Agriculture and Rural Development. Abuja. 59 p. http://fscluster.org/sites/default/files/documents/2016-nigeria-agric-sector-policy-roadmap_june-15-2016_final1.pdf.

Giller K, Tittonell P, Rufino M, Van Wijk M, Zingore S, Mapfumo P, Adjei-Nsiah S, Herrero M, Chikowo R, Corbeels M, Rowe E, Baijukya F, Mwijage A, Smith

J, Yeboah E, Van der Burg WJ, Sanogo OM, Misiko M, de Ridder N, Vanlauwe B. 2011. Communicating complexity: integrated assessment of trade-offs concerning soil fertility management within African farming systems to support innovation and development. *Agricultural Systems* 104(2):191–203. doi:10.1016/j.agsy.2010.07.002.

HarvestChoice. 2015*a*. AEZ (16-class, 2009). International Food Policy Research Institute, Washington, DC and University of Minnesota, St. Paul MN. http://harvestchoice.org/data/aez16_clas.

HarvestChoice, 2015*b*. Maize Harvested Area (ha, 2005). International Food Policy Research Institute, Washington, DC., and University of Minnesota, St. Paul, MN. http://harvestchoice.org/data/maiz_h.

Hassan Y, Abdullah AM, Ismail MM, Mohamed Z. 2014. Technical efficiency of maize production in Nigeria: parametric and non-parametric approach. *Asian Journal of Agriculture and Rural Development* 4(4):281–291.

Herbert B. 2005. Land use efficiency under maize-based cropping system in Zaria, Nigeria. *Journal of Agriculture, Forestry and the Social Sciences* 3(1):114–120.

Iheke OR, Amaechi ET. 2015. Effect of land fragmentation on smallholders' productivity in Imo State, Nigeria. *International Journal of Agricultural Science, Research and Technology in Extension and Education Systems* 5(3):195–201.

[IITA] International Institute of Tropical Agriculture. 1991. Sustainable Food Production in Sub-Saharan Africa: IITA's Contributions. International Institute for Tropical Agriculture. Ibadan, Nigeria. 195p. ISBN 978-131-086-3.

[IITA] International Institute of Tropical Agriculture. 2016. Agronomic Panel Survey for Tamasa project. Unpublished Data. International Institute of Tropical Agriculture. Ibadan, Nigeria. (www.tamasa.cimmyt.org).

Iken JE, Amusa NA. 2004. Maize research and production in Nigeria. *Africa Journal of Biotechnology* 3:302–307.

[IF] International Futures. 2019. Population data for Nigeria. IF Version 7.37 (online). Pardee Center. University of Denver, Denver. www.ifs.du.edu/ifs/frm_CountryProfile.aspx?Country=NG.

Kim SK, Fajemisin JM, Fakorede MAB, Iken JE. 1993. Maize improvement in Nigeria. Hybrid performance in the Savanna Zones. *In*: Fakorede MAB et al. (eds). *Maize Improvement, Production, and Utilization in Nigeria*. Maize Association of Nigeria. Abuja. P 41–46.

Liverpool-Tasie LSO., Omonona BT, Sanou A, Ogunleye WO. 2017. Is increasing inorganic fertilizer use for maize production in SSA a profitable proposition? Evidence from Nigeria. *Food Policy* 67:41–51.

Macauley H. 2015. Cereal Crops: Rice, Maize, Millet, Sorghum, Wheat. Background Paper – An Action Plan for African Agricultural Transformation. Feeding Africa Conference, 21–23 October 2015, Dakar, Senegal. United Nations Economic Commission for Africa. www.afdb.org/fileadmin/uploads/afdb/Documents/Events/DakAgri2015/Cereal_Crops-_Rice__Maize__Millet__Sorghum__Wheat.pdf.

Makinde AA, Bello NJ, Olasantan FO, Adebisi MA, Adeniyi HA. 2011. Seasonality and crop combination effects on growth and yield of two sorghum (*Sorghum bicolor*) cultivars in sorghum/maize/okra intercrop in a Forest-Savanna Transition Zone of Nigeria. *Agricultural Journal* 6(3):92–99.

McPherson MF. 1983. Land fragmentation in agriculture: adverse? beneficial? and for whom? Development Discussion Paper No 145. *Harvard Institute for International Development*, Harvard University. 90p.

Mustapha A, Salihu A. 2015. Determinants of technical efficiency of maize/cowpea intercropping among women farmers in Gombe State, Nigeria. *Journal of Agriculture and Sustainability* 7(2):245–258.

Nagy JG, Edun O. 2002. Assessment of Nigerian Government Fertilizer Policy and Suggested Alternative Market-friendly Policies. Report to International Fertilizer Development Center (IFDC) 69p. http://citeseerx.ist.psu.edu/viewdoc/download? doi=10.1.1.594.2361&rep=rep1&type=pdf.

Nwocha ME. 2016. Impact of the Nigerian Land Use Act on economic development in the country corporation. *Acta Universitatis Danubius. Administration* 8(2):117–128.

Nuss ET, Tanumihardjo SA. 2011. Quality protein maize for Africa: closing the protein inadequacy gap in vulnerable populations. American Society for Nutrition. *Advances in Nutrition* 2:217–224.

Obi IU. 1991. Maize: its agronomy, diseases, pests and food values. Optimal Computer Solutions Limited, Enugu, Nigeria. 208p.

Ofori F, Stern WR. 1987. Cereal-legume intercropping systems. *Advances in Agronomy* 41:41–89.

Okezie CA, Ahucgougu CU, Jamalludin S. 2012. Exploring the link between land fragmentation and agricultural productivity. *International Journal of Agriculture and Forestry* 2(1):30–34. doi:10.5923/j.ijaf.20120201.05.

Oko WE. 2011. Impact of fertilizer policy on crop production in Nigeria. MSc degree thesis submitted to the Department of Agricultural Economics, Faculty of Agriculture, University of Nigeria. 59p. www.unn.edu.ng/publications/files/images/Ozo,%20W.pdf.

Olaniyan AB. 2015. Maize: panacea for hunger in Nigeria. *African Journal of Plant Science* 9(3):155–174.

Olaniyan AB, Lucas EO. 2004. Maize hybrids cultivation in Nigeria – a review. *Journal of Food, Agriculture, & Environment* 2(3–4):177–181.

Onuk EG, Alimba JO, Kasali R. 2015. A comparative study of production efficiencies under cowpea-maize and groundnut-millet intercropping systems in the north-central zone, Nigeria. *Production Agriculture and Technology* 11(2):108–121.

Onwueme IC, Sinha TD. 1991. Field crop production in Tropical Africa – principles and practice. Center for Technical Agriculture, Ede, Netherlands. 477p. https://cgspace.cgiar.org/handle/10568/63586.

Peter B, Mungai L, Messina J, Snapp S. 2017. Nature-based agricultural solutions: scaling perennial grains across Africa. *Environmental Research* 159:283–290. doi:10.1016/j.envres.2017.08.011.

Pretty J. 2008. Agricultural sustainability: concepts, principles and evidence. *Philosophical Transaction of the Royal Society of Britain* 363:447–465.

Quainoo AK, Lawson IYD, Yawson A. 2000. Intercrop performance of maize, sorghum and soyabean in response to planting pattern. *Journal of the Ghana Science Association* 2(2):31–35.

Sanginga N, Hardarson G, Broughton WJ. 2003a. Role of biological nitrogen fixation in legume based cropping systems; a case study of West African farming systems. *Plant and Soil* 252(1):25–39.

Sanginga N, Dashiell K, Diels J, Vanlauwe B, Lyasse O, Carsky RJ, Tarawali S, Asafo-Adjei B, Menkir A, Schulz S, Singh BB, Chikoye D, Keatinge D, Ortiz R. 2003b. Sustainable resource management coupled to resilient germplasm to provide new intensive cereal – grain legume livestock systems in the dry savanna. *Agriculture, Ecosystems and Environment* 100:305–314.

Sanginga N, Okogun JA, Vanlauwe B, Diels J, Dashiell KE. 2001. Contributions of nitrogen fixation to the maintenance of soil fertility with emphasis on promiscuous soybean-based cropping systems in the moist Savanna of West Africa. *In*: Tian G, Ishida F, Keatinge JDH (eds). Sustaining Soil Fertility in West Africa. SSSA Special Publication No. 58. *Soil Science Society of America*. Madison, USA. P 157–178.

Shiferaw B, Prassana BM, Hellin J, Banziger M. 2011. Crops that feed the world 6. Past successes and future challenges to the role played by maize in global food security. *Food Security* 3:307–327.

Simmon S. 1987. Land fragmentation in developing countries; the optimal choice and policy implications. *Explorations in Economic History* 25:254–262.

Smale M, Byerlee D, Jayne TS. 2011. Maize revolutions in sub-Saharan Africa. Policy Research Working Paper 5659. Tegemeo Institute of Agricultural Policy and Development, Egerton University, Nairobi, Kenya. 47p.

Thayamini HS, Brintha I. 2010. Review on Maize Based Intercropping. *Journal of Agronomy* 9:135–145.

Uche IP. 2011. The Impact of Agricultural Policies on Nigerian Economy. MSc degree thesis submitted to the Department of Public Administration and Local Government, Faculty of Social Sciences, University of Nigeria, Nsukka. 127p. www.unn.edu.ng/publications/files/images/Uche%20Prince.pdf.

Wagner SC. 2011. Biological nitrogen fixation. *Nature Education Knowledge* 3(10):15. www.nature.com/scitable/knowledge/library/biological-nitrogen-fixation-23570419.

[WB] WorldBank. 2017. Historical population trend for Nigeria 1960–2017 (online). WorldBank Washington, DC. https://data.worldbank.org/indicator/SP.POP.TOTL?end=2017&locations=NG&start=1960.

Zohry A, Ouda S. 2017. Crop rotation could diminish summer feed gap in Egypt. *In*: Ouda S, Zohry El-Hafeez A, Noreldin T (eds). *Crop Rotation: An Approach to Secure Future Food*. P 89–109. Springer, Cham. https://doi.org10.1007/987-3-030-05351-2_6.

Zohry A E-H, Ouda S, Hamd-Alla W, Shalaby ES. 2017. Evaluation of different crop sequences for wheat and maize in sandy soil. Acta Agriculture Slovenia 109(2):383–392. http://dx.doi.org/10.14720/aas.2017.109.2.21.

8 Multifunctional land-use practices in Africa

What else do we need to do?

Elisabeth Simelton, Madelene Ostwald and Moses Osiru

Key evidence of multifunctionality from the success stories: the 'what?'

Recalling that multifunctional land use aims to produce more than one product or service, we ask: what lessons emerge from the six case studies? Let us look for a moment at the services and products produced and how farmers turned scarcities into resources.

Raising fish where there is no water

Two of the practices focus on services where water is central. Integrated watershed management is a landscape practice for better managing scarce or abundant water resources to meet several goals, such as reduced soil erosion and an increase in biomass in general (Teka Chapter 6). By reallocating water, more vegetation is sustained, and this may gradually alter the microclimate in the catchment to support a greater diversity of crops and trees, or increased crop yields. Fish farming in semi-arid environments is a realistic possibility. As technologies for recycling water advance and become affordable, it becomes a matter of selecting the appropriate fish species, identifying additional feed, and integrating with aquatic plants, fruit plants or trees, and shade-providing structures with hen houses (Matolla Chapter 5).

Recovering poor soils

Nutrient-poor soils are commonly identified as a limiting factor for African agriculture. As a collective term, climate-smart agriculture covers many kinds of practices (FAO 2013). The climate-smart agriculture examples presented here tackle multiple issues related to restoring soil carbon and soil fertility by incorporating residues and reducing tillage (Shomkegh Chapter 2). Parkland systems with scattered trees on grazing lands or on croplands produce a wide range of functions, from products like fodder, nuts, fruits or bark, to services such as improved water infiltration, shade, and carbon sequestration (Sanou Chapter 3). The shea parkland demonstrates soil-water interaction benefits between trees and associated crops. Both the cassava and the maize-based practices show that conventional staple crops such as maize and cassava, which are common in monocultures, can provide multiple benefits in diversified systems – without yield or income decline (Onoja Chapter 4; Adewopo Chapter 7). Furthermore, the integrated watershed management practice brought back groundwater tables and biomass to the landscapes (Teka Chapter 6), which has been a challenge in semi-arid landscapes where water deficits are common (Ilstedt *et al.* 2016; Nyberg *et al.* 2015).

Win-wins and triple-wins: adaptation and mitigation co-benefits

Contributions to climate change mitigation are often said not to motivate farmers to change practices and that mitigation should not be placed as an additional burden imposed on poor smallholder farmers, whose per-capita contributions to greenhouse gas emissions are miniscule. Many farming practices reported in this book represent adaptation to climate variability while contributing to increased carbon stocks in soils and vegetation (*in-situ* mitigation benefits). However, the cassava and maize cases (Onoja Chapter 4; Adewopo Chapter 7) also reveal leakage issues of agricultural expansion leading to deforestation and forest degradation elsewhere (*ex-situ* mitigation losses through emissions). In the Land Use, Land-Use Change and Forestry sector, this is one of the most contested challenges for greenhouse gas inventories. Leakage points to the importance of going beyond the fields to take a holistic view of the entire landscape with nested land uses, policy impacts at the national and international scale, and a comprehensive review of driving factors, including subsistence needs, markets, policy, and institutional factors (Duguma *et al.* 2019; Ostwald and Henders 2014). Frameworks that explore 'win–win' interactions between adaptation and mitigation and 'triple-wins' when development outcomes are added (Suckall *et al.* 2015) can guide more holistic, sustainable and hopefully long-lasting trade-off assessments. These 'win–win' interactions may not be anticipated by farmers and agriculture planners when focusing on one particular crop, practice or land use. Participatory land-use and emission scenarios can be used to simulate environmental and economic trade-offs, such as those between traditional agroforestry systems and oil palm development, to assess the policy and investment options that may enable sustainable land use (Mulia *et al.* 2013).

Land scarcity, a challenge and opportunity for multifunctional agriculture

While the term 'peri-urban agriculture' describes the location of the practice, the practice itself and its products and 'services', may vary (Onoja Chapter 4). The case study revealed that land scarcity and demand were the key factors driving the diversification of cassava-based systems. However, the role of peri-urban agriculture as a buffer of income and food for the poor should not be underestimated (Ferreira et al. 2018). With regard to land scarcity and fragmentation, the two climate-smart agriculture practices orchards and zero tillage provided an important insight, namely that smaller fields may be more cost-effective than larger ones (Shomkegh Chapter 2).

Equality benefits livelihoods

Several chapters highlight the differences between women's and men's opportunities to benefit and earn their livelihoods from agriculture (Onoja Chapter 4) and to participate in market-value chains. Examples show that women's exclusion from income-generating activities also affects other family members. The fish farming chapter illustrates how women organized themselves in groups to be stronger in market negotiations (Matolla Chapter 5). Several chapters (Shomkegh Chapter 2; Sanou Chapter 3; Teka Chapter 6) show, in various ways, that when women get involved and are able to convert 'inefficient' labour time into productive activities (with, for instance, shorter distances to water and markets) they make long-term investments. Further, the examples show that the additional incomes generated from multifunctional land uses were spent on paying back loans, on children's education, and on improved diets.

The six chapters confirm that food and ecosystem functions can be jointly produced. The multifunctionalities reported here often arose from adaptations to changes in the input supply, markets and demand, or in the natural environment. The cases contribute more diverse pictures than the conventional one of monoculture being the solution to 'feeding Africa'. Here, we emphasize that we reviewed only six cases on a vast continent that is home to countless types of land-use practices.

Processes that bring about change: the 'how?'

The chapters demonstrate multiple processes behind the transitions to more multifunctional land uses. Already in 2003 (AU 2004), African leaders had recognized that stagnant yields, poverty and food insecurity continued to hamper development throughout the continent. Through the Maputo Declaration on Agriculture and Food Security, African governments committed to allocate 10 per cent of their budgets to agriculture and rural development. This was coordinated regionally through the Comprehensive African Agricultural Development Plan (CAADP) and at national level through national agricultural investment plans aligned to CAADP goals. Then years later, a review of CAADP performance highlighted the need to set clear targets for driving agricultural development on the continent, resulting in the Malabo Declaration in 2014. They set targets such as ending hunger by 2025, halving poverty, enhancing resilience to climate change, and boosting intra-African agricultural trade. (Each country's progress on the targets can be tracked at: www.nepad.org/caadp.) International agreements like the Malabo Declaration are important mechanisms to attract investments from, for instance, the Green Climate Fund, the Global Environmental Facility, and the Bonn Challenge, to promote multifunctional practices. They provide mechanisms to ensure that knowledge, such as that highlighting benefits of multifunctional land use, can be used to support policy making at the national level.

Different contexts brought about the multifunctional land-use cases described in this book:

- **Research and government projects and interventions as enablers.** The cases with integrated watershed management and maize-based systems were driven through via government-led investments (Teka Chapter 6; Adewopo Chapter 7). This can result in scaling of interventions and meeting commitments, such as the Sustainable Development Goals and Nationally Determined Contributions, among others. While the concept of climate-smart agriculture (Shomkegh Chapter 2) at first was driven by the United Nations and members of academia, it has been advocated for and implemented through multiple stakeholder groups with guidelines in, for instance, Tanzania, Ethiopia, and Zimbabwe, and incorporated in national framework programmes, in, for example, Tanzania, Uganda, Namibia, and Kenya (Rosenstock *et al.* 2018).
- **Community action groups and advocates of practices.** 'Traditional' land uses are considered low-hanging fruit for development initiatives since support can be targeted to improve existing practices or plant improved varieties for well-tested crops and add value to existing products (Shomkegh Chapter 2; Sanou Chapter 3). This makes adoption of new practices smoother, as farmers have often already identified the problem and perhaps also the solution; and they see direct benefits of interventions (Kiptot and Franzel 2015).
- **Farmer entrepreneurs as role models who can drive changes.** The two fish farm examples show two farmers, one with more and one with fewer resources, who reached a point where they decided to take a risk and exit their comfort zones (Matolla Chapter 5). Both these farmers reached success in their risk-taking strategies. How many farmers have taken similar risks and failed, we do not know.
- **Multifunctional land use resulting from unplanned responses to changed conditions.** Earning livelihoods from staple food crops, like peri-urban cassava-based systems (Onoja Chapter 4), can be a challenge if land becomes more fragmented, land rents increase, land-use changes require investments or productivity is no longer maintained by simply adding more inputs. These conditions are in continuous change and must be monitored so that farmers and other decision-makers can

take timely action. One such change due to reduced land resources and climate change is seen in West Pokot, Kenya where pastoralists have become more sedentary over the past decades. In this process the landscape has been transformed by the establishment of enclosures made from living trees and thorny bushes, which has increased the overall biomass in the area (Nyberg *et al.* 2015). The aim with the enclosures is to separate crops from animals. Once the crops are harvested the animals are fed on the residues (Figure 8.1).

Capitalizing on benefits of multifunctional land use and research-informed policies is key

Basin-scale integrated water management combined with on-farm units for water-use efficiency has the potential for maintaining surplus water, which

Figure 8.1 Animal grazing inside enclosure after harvest. West Pokot, Kenya.
Photo credit: Ostwald 2013.

is of relevance as governments will be expected to continue to invest in large-scale water management interventions and policies (Rockström *et al.* 2010). Informed policy processes will include identifying risks, developing and testing new animal breeds, plant varieties and agronomical practices. An informed policy process will also need to document socioeconomic and environmental benefits and implications of larger-scale adoption. One such example is within the international climate policy regime and its Paris Agreement, where estimates of avoided loss and damage are most likely to be included.

Tenure is a policy area that requires attention

Governments need to pursue tenure-related issues where these restrict multifunctional land uses and land use at large. Typical situations arise when the land user is not the land-owner, and when land leases are too short to motivate long-term investments, such as permanent tree stands. Further, customary rules may, for example, forbid people of a certain gender, tribe, or economic group to use the land or be associated with particular crops or parts of crops (Kiptot and Franzel 2012; Kiptot *et al.* 2014). As a step towards resolving some of the issues associated with insecure tenure and customary law, approximately 100 countries have ratified voluntary guidelines for tenure (FAO 2012).

There is a fundamental need to understand farmers and risks

We wish to challenge common statements like 'farmers are risk-averse', 'men take more risks than women', and 'younger people take more risks than older'. What defines their comfort zone will vary from case to case. First, investing your savings in a business is different from mortgaging the land your home is on to support that business idea. Second, asymmetric information creates power imbalances, which are unlikely to benefit smallholder farmers, particularly if they are women. Relatedly, when norms exclude some groups from business arenas, the time and risks involved to first break the norms and enter those arenas (if this is even possible) are very different to those experienced by actors already on those arenas (Nyasimi and Huyer 2017). Third, farmers and land users live with risks and are on constant standby to make rapid adjustments in response to weather situations. Farmers' economic investment capacity must be seen in relation to natural disaster risks and exposures, which, in marginal and resource-poor areas, may already have depleted assets and reduced buffers for dealing with further uncertainties, risks, and stresses (Demeke *et al.* 2016). Hence, before changing a complete farming system or investing in high technology systems, it makes sense to take small steps, experiment and assess the results. The fish farming chapter (Matolla Chapter 5) illustrates the struggles and risks common to business development. On the other hand, the examples also show that diversification can become a safety net when larger investments are at stake.

Planned and unplanned actions can progress in similar ways

The lessons from the case studies show that smallholder farmers approached their innovations in different ways and that both planned and unplanned actions were shown to result in progress. We highlight four approaches with applicability and relevance regardless of location.

Small triggers that result in movement

Training can be enough to enhance both economic and environmental benefits, as seen with capacity development for women in shea processing and business skills (Sanou Chapter 3). This reminds us that smaller grants and seed funds can trigger important steps towards reaching national targets and stimulate private co-investment, including start-ups and incubator opportunities. The non-governmental sector can also achieve scale by working directly with interest groups, farmer associations and rural resource centres.

Practices can be gender neutral

Chapters by Shomkegh (Chapter 2) and Matolla (Chapter 5) show that new land-use practices can be gender neutral, and the chapter by Sanou (Chapter 3) shows that women can be empowered by targeting their traditional practices and elevating their skills in the market-value chain. By providing equal training opportunities or introducing new practices as gender neutral, each new practice is an opportunity for men and women to do things differently and avoid cementing gender roles.

Multifunctional components and practices can be shifted

One solution is to introduce a higher-value crop so that the staple shifts to being the secondary component, as in the fruit orchards in Nigeria (Shomkegh Chapter 2). Increasing the soil organic matter can enhance crop nutrient uptake in nutrient-poor soils (Aworh 2015). Most staple crops lend themselves to intercropping with legumes; improving such practices can reduce the need for fertilizers and be affordable when horticulture or perennials are not an option. Farm ponds, community managed water schemes, and solar panels bring more control and ownership to farmers (Giordano *et al.* 2018).

To avoid overconsumption of chemical agro-inputs (Shomkegh Chapter 2; Onoja Chapter 4; Adewopo Chapter 7), governments may develop guidelines for good agricultural practices. These can include certain standards for food safety, well-being of producers and environmental impacts, where part of the strategy is for some types of producers to complete certification schemes, such as Good Agricultural Practice, Rainforest Alliance Certified™ or Verified Sourcing Areas. To implement practices and benefits, communities on the ground need investments in trained extension and advisory services. Governments could fund that kind of education through Green Climate Funds.

E-farmers can build rural growth centres

The fish farm example (Matolla Chapter 5) confirms that access to longer-term credit, which farmers often demand, can benefit farm development. Mobile phone services for agriculture are advanced in many parts of Africa and India. The fact that these were hardly mentioned in the cases studied is not because of their absence but rather a sign that they are already taken for granted. Information and communication technologies bring new hope for farmers to access credit and insurance, weather forecasts and market information, to share knowledge, monitor farm activities and receive advice. Services include, for example, iCow, which sends short messages about livestock and soil management (www.icow.co.ke), mpesa in Kenya which allows farmers to access and store money using simple handsets as well as pay for services (www.safaricom.co.ke/personal/m-pesa), and Esoko, which offers agricultural advice and payment services (www.esoko.com). These kinds of services are changing the way farmers can access information previously available only to certain groups and bypass middlemen to be in direct contact with customers and more readily respond to demand. Globally, digital solutions are expected to play fundamental roles in halving total greenhouse gas emissions in all sectors, including food, transport, agriculture and forestry, by reducing food waste, planting seedlings with drones, and more efficient use of resources in precision agriculture (Falk *et al.* 2018).

The chapters on shea (Sanou Chapter 3) and fish farms (Matolla Chapter 5) show how both local jobs and businesses can be created around a multifunctional enterprise. The community knowledge centres developed around the fish farms have functions similar to social enterprises. Rural resource centres have been established in Cameroon, Burkina Faso, the Democratic Republic of Congo, Mali and Nigeria since 2006 as a community-based extension service that complements the inadequate public agricultural extension service. The centres function as training and information hubs, with a tree nursery, demonstration plots, library and meeting room facilities. They are funded through a combination of organizational support, sales and service delivery, and volunteering. The work that these rural resource centres did in responding to local needs and training farmers on tree-based systems might otherwise not have happened, given that many public extension services are underfunded (Degrande *et al.* 2015; Takoutsing *et al.* 2014).

The cases demonstrate that at certain scales, multifunctional land uses can survive without subsidies when farmers are part of the solution. When farmers and local leaders are engaged in project designs rather than passive recipients, their knowledge is respected and integrated into the solutions. Incorporating local knowledge helps in understanding how different types of land users understand and explain what happens in their environment, what matters to them, and their interactions with other groups about shared resources (Kmoch et al. 2018; Simelton and Dam 2014). For this reason, it is interesting to study top-down interventions, such as Ethiopia's watershed management programmes (Teka Chapter 6), which resemble those in China and Vietnam in the 1990s and 2000s (Bachewe et al. 2018). Here, large-scale interventions in extension and availability of inputs (especially financial), combined with farmers' contributions of labour-for-food, seem to have worked, in times when and places where economic development standards were quite similar. Planners will now need to avoid creating new problems when solving an environmental issue.

Benefits of multifunctional land uses for Africa: the 'So what?'

While many seem to agree that we need to increase yields and ensure diverse diets, research on food security seems to focus on either the quantity or the quality of food – and smallholder farmers are often forgotten either way (Ickowitz et al. 2019). In this book, we have tried to show a variety of agricultural practices that return more than the yields to smallholders' livelihoods and communities. Returning to Wiggering et al. (2006) in Simelton, Ostwald and Osiru (Chapter 1), we ask: knowing all these benefits, so what? Which of the values of multifunctional land use does the rest of society perceive to be important so that these environmental and social functions can be maintained?

The costs of poverty and food insecurity

Some of the case studies showed what happens when poor households increase their incomes. They invested in short- and long-term returns: improved diets and their children's education. While counting the number of poor is comparatively straightforward, estimating the cost of poverty is more complex. A study from the United States showed that the cost of child poverty is about 5.4 per cent of the gross domestic product and estimated that every dollar the country spent on reducing childhood poverty would save at least seven dollars (McLaughlin and Rank 2018). In developing countries, it turns out that the net food and agricultural exporters invested more in social protection programmes that benefitted the rural poor, than those with agricultural trade deficits and manufacturing trade surpluses (Desai and Rudra 2018).

The public costs for food insecurity, such as those of the civil unrest and recovery of people who fell into poverty with the food price inflation 2007–2008 (Berazneva and Lee 2013; Simelton 2010, 2011; Veninga and Ihle 2018), are not explicit in the frameworks of Garibaldi et al. (2017) and Vereijken (2003). These may depend on European-centred frameworks that take institutional roles and food security for granted and instead aim to embed natural, cultural and recreational values in schemes for payments for ecosystem services. In some cases, the roles for multifunctional agriculture and smallholders are clearly stressed, such as Niger's socio-economic development plan 2012–2015 (FAO 2015). Others argue that global food prices are linked to the food security of urban low-income net food buyers. In such contexts, commercial medium-scale farms are expected to contribute to food security by job creation and (rural) wages (Meyfroidt 2018). Furthermore, the benefits of agricultural exports are less clear when it comes to foreign acquisition of agriculture land. A global estimate of large-scale land acquisition for commercial agriculture shows that although the relevant area could feed 300 to 500 million people through intensification, the food is exported from countries with a high prevalence of poverty and malnourishment, disrupting their sources of livelihoods (Rulli and D'Odorico 2014).

The yield gap

Rainfed agriculture continues to play an important role for many farming systems, while yields in many countries are less than 30–40 per cent of their potential yields (Rockström et al. 2010). Feeding the growing population on less farmland will require a transformation of the whole agrifood system as we know it. The value chain starts with improved stress-tolerant seeds and a combination of diverse farming systems that are adapted to new climatic situations. Some of the potentially climate resilient crops suitable for Africa are generally under-researched, such as pigeon pea, cowpea, sweet potato, lentils, and chickpeas (Manners and van Etten 2018). New business opportunities may arise from taking advantage of underutilized food crops, so-called orphan crops, for enhanced nutritional diets (Aworh 2015, see also http://africanorphancrops.org/), and exploring the abundance of wild foods that can be domesticated (Bvenura and Sivakumar 2017). Integrated watershed management combined with climate-smart agriculture and fish farming (Shomkegh Chapter 2; Matolla Chapter 5; Teka Chapter 6) interventions are promising examples for the potential reduction of yield gaps. The example from the Tigray region of Ethiopia estimated investment costs for water harvesting bunds in the three watersheds to be between US$29 and 87 per hectare, with annual maintenance costs of US$1.7 and 6.1 per hectare (Teka Chapter 6). In industrial production, such expenses are added to the consumer price. Here, weighted against gains in food security, people's living standards and re-greening of landscapes – what is a 'fair price'? How should a 'fair price' be defined?

Local and global values of agriculture ecosystems

Who needs to pay greater attention to intensification and expansion of agriculture? The literature reviewed by Garibaldi *et al.* (2017) compared 154 conventional and 13 alternative practices, such as sustainable intensification, organic, diversification, ecological intensification and agroecological farming systems. Interestingly, 61 per cent of the comparisons showed greater crop yield for alternative rather than conventional practices, whereas about 20 per cent showed the opposite trend and another 20 per cent showed no differences. Similarly, two-thirds of the comparisons achieved greater farm profitability for alternative practices, while 11 per cent found the opposite trend, and 23 per cent showed no differences. Few of the studies provided quantitative data on both crop yield and socioeconomic indicators, such as well-being; hence, little evidence was documented on the multifunctionality of alternative practices. Furthermore, when agriculture intensification involves conversion of forests and grassland to agriculture, this poses threats to natural resources and habitats. National and subnational decision-makers can develop policies that reduce land conversions while building up habitat quality on existing agricultural land. Policies can also be designed to give farmers incentives to invest in conservation agriculture and agrobiodiversity, including by offering tenure security and access to credit and efficient markets (Perrings and Halkos 2015). Furthermore, countries can prevent agriculture-driven deforestation, with or without large-scale land acquisitions, by including land management principles, for example, in their REDD+ strategies (Carter *et al.* 2017). The 'Economics of Ecosystems and Biodiversity AgriFood' initiative is a multidisciplinary platform that provides guidance for more comprehensive evaluations of eco-agri-food systems (http://teebweb.org/agrifood/).

Resilience to environmental degradation and climate change impacts

Integration of more trees in agriculture and farming practices that prevent land degradation can enhance carbon sinks (Zomer *et al.* 2016). A suitability mapping of shea trees shows a potential distribution on 340 million hectares across 23 countries (Naughton *et al.* 2015). The mapping study estimated that this corresponds to 1.8 billion trees and would involve 18 million women collectors. As technology improves, remote-sensing tools will enable us to count individual trees. Using remote sensing, Bastin *et al.* (2017) identified 467 million hectares of dryland forests that had not been reported previously. Further, they estimated that 1,327 million hectares of drylands had more than 10 per cent tree cover in 2015, such as the shea parklands. Such remote-sensing methods can offer affordable and objective solutions for monitoring tree plantation efforts and estimating their benefits, which are often among the most difficult parts of reporting on commitments to the United Nations Framework Convention on Climate Change and carbon financing projects (Rosenstock *et al.* 2018).

The cost of adapting or not adapting agriculture to climate change

Estimating and comparing the costs and benefits of different adaptation options, including not adapting, is a complex matter that depends on the type of calculated and emerging risks and the projected frequency and intensity of those risks (Klein *et al.* 2014). It also involves consideration of the ethically acceptable risks and adaptation opportunities among different groups of individuals (Niang *et al.* 2014). Estimates suggest that the cost of not adapting farming systems to climate change will be about 5 per cent of the gross national product by 2030, while estimates of adaptation costs range from two US dollars per person for a national climate change strategy in Rwanda to six US dollars per person for protecting pastoralist and livestock systems in Tanzania. More importantly, delayed action was estimated to cost ten times more by 2030 (IIED 2011). It is becoming more evident that public sources will become insufficient and that private finance is needed. Climate finance from public sources is typically given to profitable mitigation interventions, for example renewable energy, rather than to adaptation activities in the land-use sector (Oliver *et al.* 2018). In 2017, private climate finance at global level was reported at 249 billion US dollars. Of this amount, 238 billion was for renewable energy (Oliver *et al.* 2018). There are opportunities to include multifunctional agricultural land as part of green infrastructure and ecosystem-based adaptation strategies in adaptation funds or payment for ecosystem services schemes, where consumers recognize environmental services achieved by farmers. The buffering roles of agriculture during environmental and economic crises must be reflected in budget allocations for disaster risk and climate adaptation (FAO 2018).

The role of science in promoting sustainable land-use practices and food security: 'what else do we need to know?'

The examples from this book show that the proper quantification and valuation of multiple products and services from land has room for scientific and practical improvements. Documenting the multiple functions that multifunctional land uses have will include their resilience to external stress, the value of replacing external inputs with ecosystem services, and complementarity or positive interactions. Garibaldi *et al.* (2017) suggested an evidence framework that draws on social, human, cultural, natural, financial, and economic assets (Table 8.1). Arguing that food security will

Table 8.1 Example of indicators for assessing the multifunctionality of farming practices

Rural functions Vereijken 2003	Typical performance indicators Garibaldi et al. 2017	Our suggested indicators Research questions, partly drawn from this book, to compare 'conventional' and multifunctional land use
Health and wellness	Encourages non-farmed species diversity	• Does starting the practice require a particular 'farmer characteristic': are food security, start-up capital/time, certain social/human assets prerequisites for the practice? • How does the land use contribute to household/local/national food security? • Does the land use cement existing social or gender inequalities? Does the practice help free up unpaid time or reduce physically demanding workloads or exposure to dangerous substances?
Nature and landscape	Encourages spatial heterogeneity Explicit focus on traditional knowledge	• What biodiversity values are enhanced by the land use? • Does the land use help beautify the landscape or contribute to agro-tourism? • Is the research design informed by various local groups' knowledge and needs from the beginning?
Environment and climate	Use of synthetic inputs vs use of organic inputs Exploits ecosystem services	• What are the values of replacing external inputs with ecosystem services (for instance, exchanging pesticides for biological pest control, inorganic fertilisers for compost), or complementarity and positive interactions? • To what extent are herbicides solely used to save labour costs for weeding? • Does the land use reduce sensitivity to adverse climatic impacts? Does the land use contribute to sequestering carbon or reducing greenhouse gas emission, directly or indirectly (through its value chain, such as reducing transports)?
Production	Uses diverse crop/livestock species	• How is land-use efficiency evaluated? What are the individual and combined differences in yield and income?

Table 8.1 Continued

Rural functions Vereijken 2003	Typical performance indicators Garibaldi et al. 2017	Our suggested indicators Research questions, partly drawn from this book, to compare 'conventional' and multifunctional land use
Wealth and income	Highly labour dependent	• Are new jobs created? Do they develop new specialists or service providers, such as processing, information and communication technologies, intermediaries, transport? • Do multifunctional farming systems avoid market saturation and a rural economy dependent on few products?
NA	Plans for resilience Exploits processes at multiple temporal and spatial scales	• What are the costs, benefits, and potential risks associated with the land use? How are they balanced? How do benefits spill over to the wider community and natural environment? • Who is looking for return on investment (public or private, grants or loans) and over what period? **Impacts at scale** • Social and institutional buy-in in top-down interventions versus the role of markets in driving multifunctional practices. • What policies are in place that enable or discourage multifunctional practices? What policymakers and other actors are involved/missing as stakeholders in the process? • Are rural and urban food-security links strengthened?

Sources: Adapted from Vereijken (2003) and Garibaldi *et al.* (2017).

not be solved by increasing crop yields alone, they take agricultural sustainability to depend on government and civil society actions, including rural communities, researchers and technicians. Here, we have modified the frameworks of Vereijken (2003) and Garibaldi *et al.* (2017) to make explicit institutional functions and food security (meaning quantitative and qualitative stability of nutrients) and stress the impacts at scale. Similar classifications have been adapted for modelling tools, such as the Common

148 *Elisabeth Simelton* et al.

International Classification of Ecosystem Services (https://cices.eu/supporting-functions/; Potschin and Haines-Young 2011), and developed into typologies for mapping flows of ecosystem functions (Pagella and Sinclair 2014).

To get a sense of the role of science in African agriculture, we conducted a small anonymous survey among different scientific and agricultural networks globally with the aim of giving us an indication of relevant issues. Without any goal of methodological soundness or genuine analysis, this provides a hint of some of the debates that exist today.

The first question in the survey was: 'What is the major critical research area that needs to be in focus to sustainably strengthen African agriculture?' Two major research needs appeared: (i) adaptation to change, (ii) climatic and entrepreneurial and market-related knowledge. In the former case, respondents mentioned modalities of agricultural management in a changing climate and, more explicitly, in the event of shocks. The source of information given to smallholder farmers was stated as problematic and fragmented or too homogenous. On the same note, there is a lack of knowledge on the environmental impacts of different types of production systems. Respondents also highlighted the need for more knowhow on the value chains of agricultural products and on how farmers can enter the market and become entrepreneurial actors in the African agriculture sector. Based on our own non-scientific interpretation, the narratives that are being retold regarding African agriculture are hampering the development of the same. This dominating, repetitive and unfavourable narrative is also the basic idea that has been driving this book project.

The second question we asked was about the 'most damaging myths about African agriculture'. Some respondents stated that 'farming equals poverty', 'African agriculture is one homogenous system', and 'soil degradation is irreversible' as examples of such myths. The dominating myth damaging African agriculture, however, relates to 'the irrational African farmer', 'the inefficient production', and 'that farmers are not forward-looking or market-oriented'. Although it is possible to find scientific evidence to support each of these statements, the myths are created when one repeated narrative points to an immutable nature of African agriculture. Rather than fuelling damaging myths, the role of science is to contribute more diverse realities and bring constructive evidence of ongoing agriculture developments in Africa that are taking farmers, consumers and leaders on long-term sustainable trajectories.

This brings us to the third question: 'How is the myth, true or untrue, hampering sustainable development of agriculture in Africa?' The answers from our colleagues pinpointed this drawback by exemplifying how these narratives or myths drive general policy processes in Africa. For example, efforts that focus on developing new agricultural technologies at a fairly scientific and technocratic level rather than on the adoption of technologies that already exist, a process strongly driven by how funding streams flow into agricultural research and development. One approach, with lower investment cost and faster adoption, could be to build on existing and well-functioning technologies that can reduce those risks. These practices are often sporadic and contextual, and therefore not well known, well documented or well presented. Another example is the impact that unfavourable myths have on youth in agriculture, since 'farming is portrayed as a non-prosperous or bad career choice', making it an unattractive option for young people, which on the other hand is not unique to Africa. The remedy to this downward spiral is to showcase that money can be earned in agriculture and that it can offer a good livelihood. Enhancing the appeal of agriculture requires investments in infrastructure, including roads, markets, rural services, and irrigation, and clear incentives for adopting new technologies and becoming more involved in post-harvest processing stages of the value chain. 'Abandoning the one-size-fits-all solution within extension and policy' and 'focusing on enhancing agricultural and context-specific research' were suggested as steps on the path towards more prosperous agricultural progress.

Finally, we asked our colleagues to think of 2063, linking to the Africa Union Agenda 2063 for the socioeconomic transformation within the African Union (AUC 2015). We asked them to 'state the biggest risks and strengths within African agriculture'. The three major risks they foresee are (i) impacts of climate change and associated water stress, (ii) the looming population increase, and (iii) land shortage. Three strengths were seen in (i) African youth who are expected to be better educated than today, (ii) the richness in natural resources, such as favourable growing climates and minerals, and (iii) diversity of products, production systems and market channels that hold great potential.

Where do we go from here?

Scientists have raised concerns over the promotion of single adaptation responses – such as crop insurance or new crop varieties – that increase the vulnerability to climate risks by disincentivizing practices that would lead to more positive outcomes over longer time scales. Vermeulen *et al.* (2018) reviewed case studies that met their criteria for transformational adaptation to climate change, including eight African agricultural systems. Among the successful transformational changes in Niger, was, not just giving farmers technical assistance, but also control over assets. The study concludes that governments and development partners could improve the effectiveness of outcomes by providing more comprehensive and long-term approaches to adaptation planning alongside financial and technical assistance, within a framework that rewards farms as multifunctional systems.

This involves a shift from the global to local levels, to understand and economically reward farms as multifunctional land-use systems that deliver food (health and nutrition), profits, jobs, environmental benefits and cultural value that goes beyond national food security. The role of governance is to ensure inclusive decision-making and distribution of outcomes. Adaptation processes need to be implicitly included within the Comprehensive Africa Agriculture Development Programme (CAADP), which is Africa's framework for agricultural transformation reinforced by the 2014 Malabo Declaration, the Science, Technology and Innovation Strategy for Africa 2024, commitments to the United Nations Framework Convention on Climate Change such as the global stocktake, Nationally Determined Contributions through the Green Climate Fund, and loans and grants from development banks. Technical and financial assistance for identifying adaptation options may include compensation for transformative changes, information, and knowledge systems that give farmers tools to forecast possible futures, and for monitoring systems that give early warning of agricultural systems being on the wrong track, away from long-term sustainability (Niang *et al.* 2014; Vermeulen *et al.* 2018). The importance of these issues needs to be highlighted in the curriculum for future leaders.

The six case studies presented in this book provide promising alternatives to the conventional view that global food security requires large-scale monoculture production of staple crops. Research on multifunctional land use can help us better understand the interactions in these diverse socioecological systems.

Our cases have mainly concerned multifunctional practices that may be incremental adaptation responses to current risks; in particular, water, rainfall, and food and land security and shortages. Identifying various factors as aspects of past and current success does not mean they would enable near-term or long-term future sustainability; in general, there are temporal trade-offs between short- and long-term goals or spatial trade-offs, for example between ending some land use now for the sake of setting aside land elsewhere.

We hope that this book will inspire, provoke reflection and action on enhanced multifunctional land use, and initiate more research.

References

[AU] African Union. 2004. The Comprehensive African Agricultural Development Program – CAADP. African Union. Addis Ababa, Ethiopia. www.nepad.org/caadp.

[AUC] African Union Commission. 2015. Agenda 2063 The Africa We Want. 01 Background Note. African Union Commission. Addis Ababa. https://au.int/sites/default/files/documents/33126-doc-01_background_note.pdf.

Aworh O. 2015. Promoting food security and enhancing Nigeria's small farmers' income through value-added processing of lesser-known and under-utilized indigenous fruits and vegetables. *Food Research International* 76(4):986–991.

Bachewe FN, Berhane G, Minten B, Taffesse AS. 2018. Agricultural Transformation in Africa? Assessing the Evidence in Ethiopia. *World Development* 105:286–298.

Bastin J, Berrahmouni N, Grainger A, Maniatis D, Mollicone D, Moore R, Patriarca C, Picard N, Sparrow B, Abraham E et al. 2017. The extent of forest in dryland biomes. *Science* 356:635–638.

Berazneva J, Lee D. 2013. Explaining the African food riots of 2007–2008: An empirical analysis. *Food Policy* 39:28–39.

Bvenura C, Sivakumar D. 2017. The role of wild fruits and vegetables in delivering a balanced and healthy diet. *Food Research International* 99(1):15–30.

Carter S, Manceur A, Seppelt R, Hermans-Neumann K, Herold M, Verchot L. 2017. Large scale land acquisitions and REDD+: a synthesis of conflicts and opportunities. *Environmental Research Letters* 12(3):035010.

Degrande A, Tchoundjeu Z, Kwidja R, Fouepe G. 2015. Rural Resource Centres: a community approach to agricultural extension. Note 10. *GFRAS Good Practice Notes for Extension and Advisory Services.* Lindau, Switzerland. www.world agroforestry.org/downloads/Publications/PDFS/BR15624.pdf.

Demeke M, Kiermeier M, Sow M, Antonaci L. 2016. Agriculture and food insecurity risk management in Africa. Concepts, lessons learned and review guidelines. Food and Agriculture Organization of the United Nations. Rome. 92p. www.fao.org/3/a-i5936e.pdf.

Desai R, Rudra N. 2018. Trade, poverty, and social protection in developing countries. *European Journal of Political Economy.* doi.org/10.1016/j.ejpoleco.2018.08.008.

Duguma L, Atela J, Minang P, Ayana A, Gizachew B, Nzyoka J, Bernard F. 2019. Deforestation and forest degradation as an environmental behavior: unpacking realities shaping community actions. *Land* 8(2):26.

Falk J, Gaffney O, Bhowmik AK, Borgström-Hansson C, Pountney C, Lundén D, Pihl E, Malmodin J, Lenhart J, Jónás K et al. 2018. Exponential climate action roadmap future Earth. Sweden. 108p. https://exponentialroadmap.org/wp-content/uploads/2018/09/Exponential-Climate-Action-Roadmap-September-2018.pdf.

[FAO] Food and Agriculture Organization of the United Nations. 2012. Voluntary Guidelines on the Responsible Governance of Tenure of Land, Fisheries and Forests in the Context of National Food Security. Food and Agriculture Organization of the United Nations. Rome. www.fao.org/docrep/016/i2801e/i2801e.pdf.

[FAO] Food and Agriculture Organization of the United Nations. 2013. Climate-smart Agriculture – Sourcebook. Food and Agriculture Organization of the United Nations. Rome. 570p. www.fao.org/3/i3325e/i3325e.pdf.

[FAO] Food and Agriculture Organization of the United Nations. 2015. Collaboration for Strengthening Resilience. Country Case Study Niger. Food and Agriculture Organization of the United Nations, International Fund for Agricultural Development, World Food Programme. Rome. 20p. https://documents.wfp.org/stellent/groups/public/documents/newsroom/wfp278361.pdf?_ga=2.249800582.1647555720.1549691044-1411497054.1549691044.

[FAO] Food and Agriculture Organization of the United Nations. 2018. 2017 The Impact of Disasters and Crises on Agriculture and Food Security. Food and Agriculture Organization of United Nations. Rome. 168p. www.fao.org/policy-support/policy-themes/disaster-risk-reduction-agriculture/en/.

Ferreira A, Guilherme R, Ferreira C, Oliveira M. 2018. Urban agriculture, a tool towards more resilient urban communities? *Current Opinion in Environmental Science and Health* 5:93–97.

Garibaldi LA, Gemmill-Herren B, D'Annolfo R, Graeub B, Cunningham S, Breeze T. 2017. Farming Approaches for Greater Biodiversity, Livelihoods, and Food Security. *Trends in Ecology & Evolution* 32(1):68–80. doi:10.1016/j.tree.2016.10.001.

Giordano M, Barron J, Unver O. 2018. Water scarcity and challenges for smallholder agriculture. *In*: Campanhola C, Pandey S (eds). *Sustainable Food and Agriculture. An integrated approach*. Academic Press. Cambridge, USA. https://hdl.handle.net/10568/988597.

Ickowitz A, Powell B, Rowland D, Jones A, Sunderland T. 2019. Agricultural intensification, dietary diversity, and markets in the global food security narrative. *Global Food Security* 20:9–16.

[IIED] Institute for Environment and Development. 2011. Costing agriculture's adaptation to climate change. The International Institute for Environment and Development. London. 4p. http://pubs.iied.org/pdfs/17120IIED.pdf.

Ilstedt U, Bargués A, Tobella HRB, Bayala J, Verbeeten E, Nyberg G, Sanou J, Benegas L, Murdiyarso D, Laudon H, Sheil D, Malmer A. 2016. Intermediate tree cover can maximize groundwater recharge in the seasonally dry tropics. *Nature Scientific Reports* 6:21930.

Kiptot E, Franzel S. 2012. Gender and agroforestry in Africa: a review of women's participation. *Agroforestry Systems* 84(1):35–58.

Kiptot E, Franzel S. 2015. Farmer-to-farmer extension: opportunities for enhancing performance of volunteer farmer trainers in Kenya. *Development in Practice* 25(4):503–517.

Kiptot E, Franzel S, Degrande A. 2014. Gender, agroforestry and food security in Africa. *Current Opinion in Environmental Sustainability* 6:104–109.

Klein RJT, Midgley GF, Preston BL, Alam M, Berkhout FGH, Dow K, Shaw M. 2014. Adaptation opportunities, constraints, and limits. *In*: Field CB, Barros VR, Dokken DJ, Mach KJ, Mastrandrea MD, Bilir TE, Chatterjee M, Ebi KL, Estrada YO, Genova RC et al. *Climate Change 2014: Impacts, Adaptation, and Vulnerability. Part A: Global and Sectoral Aspects*. Contribution of Working Group II to the Fifth Assessment Report of the Intergovernmental Panel on Climate Change. Cambridge University Press, Cambridge, United Kingdom and New York, NY, USA. P 899–943.

Kmoch L, Pagella T, Palm M, Sinclair F. 2018. Using local agroecological knowledge in climate change adaptation: a study of tree-based options in Northern Morocco. *Sustainability* 10(10):3719.

Manners R, van Etten J. 2018. Are agricultural researchers working on the right crops to enable food and nutrition security under future climates? *Global Environmental Change* 53:182–194.

McLaughlin M, Rank M. 2018. Estimating the economic cost of childhood poverty in the United States. *Social Work Research* 42(2):73–83.

Meyfroidt P. 2018. Trade-offs between environment and livelihoods: bridging the global land use and food security discussions. *Global Food Security* 16:9–16.

Mulia RA, Widayati A, Suyanto S, Agung P, Zulkarnain M. 2013. Low carbon emission development strategies for Jambi, Indonesia: simulation and trade-off analysis using the FALLOW model. *Mitigation and Adaptation Strategies for Global Change*. doi:10.1007/s11027-013-9485-8.

Naughton C, Lovett P, Milhelcic J. 2015. Land suitability modeling of shea (Vitellaria paradoxa) distribution across sub-Saharan Africa. *Applied Geography* 58:217–227.

Niang I, Ruppel O, Abdrabo M, Essel A, Lennard C, Padgham J, Urquhart P. 2014. Africa. In: Barros VR, Field CB, Dokken DJ, Mastrandrea MD, Mach KJ, Bilir TE, Chatterjee M, Ebi KL, Estrada YO, Genova RC et al. (eds) *Climate Change 2014: Impacts, Adaptation, and Vulnerability. Part B: Regional Aspects.* Contribution of Working Group II to the Fifth Assessment Report of the Intergovernmental Panel on Climate Change. Cambridge University Press, Cambridge, United Kingdom and New York NY. P 1199–1265.

Nyasimi M, Huyer S. 2017. Closing the gender gap in agriculture under climate change. *Agriculture for Development* 30:37–40.

Nyberg G, Knutsson P, Ostwald M, Oborn I, Wredle E, Otieno D, Mureithi S, Mwangi P, Said M, Jirström M et al. 2015. Enclosures in West Pokot, Kenya: transforming land, livestock and livelihoods in drylands. *Pastoralism: Research, Policy and Practice* 5:25.

Oliver P, Clark A, Meattle C. 2018. Global Climate Finance: An Updated View 2018 Climate Policy Initiative. https://climatepolicyinitiative.org/publication/global-climate-finance-an-updated-view-2018/; https://climatepolicyinitiative.org/wp-content/uploads/2018/11/Global-Climate-Finance-An-Updated-View-2018.pdf.

Ostwald M, Henders S. 2014. Making two parallel land-use sector debates meet: carbon leakage and indirect land-use change. *Land Use Policy* 36:533–542.

Pagella T, Sinclair F. 2014. Development and use of a typology of mapping tools to assess their fitness for supporting management of ecosystem service provision. *Landscape Ecology* 29(3):383–399.

Perrings C, Halkos G. 2015. Agriculture and the threat to biodiversity in sub-Saharan Africa. *Environmental Research Letters* 10(9). doi:10.1088/1748–9326/10/9/095015.

Potschin M, Haines-Young R. 2011. Ecosystem services: Exploring a geographical perspective. *Progress in Physical Geography* 35(5): 575-594.Rockström J, Karlberg L, Wani S, Barron J, Hatibu N, Oweis T, Bruggeman A, Farahani J, Qiang Z. 2010. Managing water in rainfed agriculture – the need for a paradigm shift. *Agricultural Water Management* 97:543–550.

Rosenstock T, Wilkes A, Jallo C, Namoi N, Bulusu M, Suber M, Bernard F, Mboi D. 2018. Making trees count: measurement, reporting and verification of agroforestry under the UNFCCC. CGIAR Research Program on Climate Change, Agriculture and Food Security (CCAFS). Wageningen, the Netherlands. https://hdl.handle.net/10568/98404.

Rulli M, D'Odorico P. 2014. Food appropriation through large scale land acquisitions. *Environmental Research Letters* 9:064030.

Simelton E. 2010. Don't we all want good weather and cheap food? In: Dodson J (ed.). *Changing Climates, Earth Systems and Society.* Springer. P 201–215.

Simelton E. 2011. Food self-sufficiency and natural hazards in China. *Food Security* 3:35–52.

Simelton E, Dam VB. 2014. Farmers in NE Viet Nam rank values of ecosystems from seven land uses. *Ecosystem Services* 9:133–138.

Suckall N, Stringer LC, Tompkins EL. 2015. Presenting triple-wins? Assessing projects that deliver adaptation, mitigation and development co-benefits in rural Sub-Saharan Africa. *Ambio* 44(1):34–41.

Takoutsing B, Tchoundjeu Z, Degrande A, Asaah E, Tsobeng A. 2014. Scaling-up sustainable land management practices through the concept of the rural resource centre: reconciling farmers' interests with research agendas. *International Journal of Agricultural Extension Education* 20(5):463–483.

Veninga W, Ihle R. 2018. Import vulnerability in the Middle East: effects of the Arab spring on Egyptian wheat trade. *Food Security* 10(1):183–194.

Vereijken P. 2003. Transition to multifunctional land use and agriculture. *NJAS Wageningen Journal of Life Sciences* 50(2):171–179.

Vermeulen S, Dinesh D, Howden S, Cramer L, Thornton P. 2018. Transformation in practice: a review of empirical cases of transformational adaptation in agriculture under climate change. *Frontiers in Sustainable Food Systems*. https://doi.org/10.3389/fsufs.2018.00065.

Wiggering H, Dalchow C, Glemnitz M, Helming K, Müller K, Schultz A, Stachow U, Zander P. 2006. Indicators for multifunctional land use – linking socio-economic requirements with landscape potentials. *Ecological Indicators* 6(1):238–249.

Zomer R, Neufeldt H, Xu J, Ahrends A, Bossio D, Trabucco A, van Noordwijk M, Wang M. 2016. Global tree cover and biomass carbon on agricultural land: the contribution of agroforestry to global and national carbon budgets. *Nature Scientific Reports* 6. doi:10.1038/srep29987.

Index

Page numbers in **bold** denote tables, those in *italics* denote figures.

acacias 27, 47; *Acacia nilotica* 27; *Acacia raddiana* 47; *Acacia Senegal* 47
Adansonia digitata 47
adaptation (s) 11; benefits 10; to change 148; to changes in input supply 136; climate 7, 38, 145; low-cost 16; and mitigation co-benefits 135; passed on from generation to generation 3; responses, single 149; scenario of no 22; of traditional systems 16
adaptation (s) actions/activities in land-use sector 38, 145; costs 145; options/opportunities 145, 150; planning 149; responses, incremental 150
adaptation strategy/strategies 38; identification of 40; ecosystem-based 145; National Strategy 38
agricultural policy/policies 128; Agricultural Promotion 39, 73, 119, 127; on climate-smart practices 38; document for promotion 117; European Common 1; framework to promote 109; to give farmers incentives 144; national 74; none promoting multiple-cropping systems 127; providing support to encourage farming 119; support lacking 74
Agricultural Transformation Agenda 119
agrobiodiversity benefits 40; incentives for farmers to invest in 144
agroforestry 3, 8, 29, 37, **63**; in climate-smart village programme 23; estimated benefits from 38; farmer-managed natural regeneration approach 54; in least developed African countries 7; in NDCs, account for 7, 10–11; promoting 38; reforestation stage 11; in the Sahel 53; strategies 6; traditional systems 135; urban and peri-urban 7
agroforestry parklands 47, 49–50, 52, 55–6; multifunctional landscapes 47; reduction in tree density on 55; systems 47, 53, 56; trees under forest legislation 56; West African 47
agroforestry parklands, shea production in 49–50; stakeholders in 52
animal 56; breeds improved 109; breeds, testing of new 139; domestic farm 24; health and well-being 2; husbandry 98; kept close to trees 3; local herbal medicines for 106; separated from crops by enclosures *138*; sustainable productivity enhanced 39; traction 98; wild, hunting 64; *see also* livestock
animal feed: cassava roots for 62; core ingredient maize 115; crop residues collected for 50; grains for 50; pasture 39; peri-farmers use income from cassava to buy 70; sufficiency 22
annual crops: and agroforestry **63**; on land with insecure tenure 25; lower yields from 53; main livelihoods from 98; rotation of **65**
aquaculture 89; based land use 84; gender division of labour 83; highland-based 88; training on techniques 83; use of nutrient-loaded water for irrigation 88; women-dominated 83

Index

aquaculture, African 78; contribution to food security 91; development 79, 81; freshwater 78, 80; not recognized as businesses 90; provides ecosystem goods and services 79; socioeconomic benefits 78

Aquaculture Collaborative Research Support Program 79

aquaculture-horticulture farmers, integrated 90; systems 91

aquaculture in Kenya: advisory services 80; annual production increased 79, 80; The Blue Growth Initiative 92; development 79; Economic Stimulus Program for 82; government interventions suggested 91; integrated intensification 81; Kamuthanga Fish Farm 89; policies, current 81; role of gender in 82; semi-intensive systems 81; sustainable development 81, 91–2; technologies, awareness about 92; viable investments 92; women employed in 83

aquaculture systems: integrated 79, 111; intensive 89; recirculating 86, 87, 88–9

Asia 4, 10, 14

bee (s) 50; hives 50, 106; keeping 50, 98, 106; production, honey 106

BIODEV 53

biodiversity 2; agro-biodiversity benefits 40; banks 5; commitments on 6; conservation 11, 22; Conventions on (CBD) 7; loss 23, 36; values **146**

Blue Growth Initiative, The 92

Brazilian Low-Carbon Agriculture Plan 7

Burkina Faso 16, 54; CCAFS programme 23; ecozones 47; gendered traditions turned into opportunities 15; rural landscapes uses 48; rural population tripled 55; rural resource centres established 141; SAFRUIT project 53; Sudanese zone land-use system 49

Burkina Faso, shea trees: belong to land owner 10; African mistletoe species infestation 55; butter processed and traded 51; caterpillars 50; consumption of butter 50; intercropping of 27, 49; listed as protected species 56; nuts 52; nuts collected and processed by women 52–3; production 3, 9; variability of nut yield 55

canopy/canopies 8, 12, 50; multi-level structures 125; multiple layers 3; open/closed 27; planting shade-tolerant crops beneath 48

carbon 11; credits 12; emission reductions, lifetime 38; financing projects 145; pool of multifunctional land uses 10; storage 1–2, 12; *see also* Brazilian Low-Carbon Agriculture Plan

carbon sequestration 29, 134; farming practices that contribute to 10; land use contribution to **146**; potential 9; by shea trees 51

carbon in soil: improved by presence of shrubs and trees 47; increase 29; restoring 134; retained 40; sinks enhanced by integration of more trees 144

carbon stock: declining 9; farming practices contributing to increased 135; global contribution 8; improvements through agroforestry, forest management and tree planting 53

cash: crop production for food security 96; crops 49–50; incentives provided by programmes 108–9; income increasing 106; incomes 9; limited access to 126; poorer households needing to work extra for 126; return increased **123**

caterpillars, shea 50

cattle 24, 50, 98; trampling by 39

cereals 34; based diets 48; declining trend of national agricultural productivity in 118; harvesting and processing technologies 36; integration/intercropping with 47, 50, 122, 127; processed 115; water from aquaculture used to irrigate 88

cereal yields decline 22; increased 48; interfered with by non-productive trees 55; shade main cause of decrease 48

certification: of farm operations for safe fish 91; land-tenure 110–11; schemes 8, 141

CGIAR Research Program on Climate Change, Agriculture and Food Security (CCAFS) 23

Index   157

children: additional incomes spent on education for 136, 142; benefitted from irrigation facilities 106; carry water for domestic purposes 89; farmers' income gains invested in education for 74, 89; helped move soil and rock to restore landscapes 101; important source of protein for 50; income-generating activities contributed to education for 111; multiple-cropping systems, implications for **125**; Nigerian, chronically malnourished 22; opportunities to attend school 126; pond management activities carried out by 83

climate 2, **146**; buffering 47; crop model studies 12; driven plan 7; favourable growing 149; finance from public sources 145; Fund, Green 7, 137, 141, 150; impact-food security studies 11; information 37–8; mitigation 2, 6–7; policy regime, international 139; properties of a landscape 13; resilience, building 71; resilient crops 143; semi-arid 84, 98; variability 56, 96, 135

climate adaptation 7; actions in agriculture 38; budget allocations for 145

climate change 16; adaptation, best practices for 39; adverse impacts from 96; CGIAR Research Program on 23; combatting 23; costs of national adaptation strategy 145; enhancing crop productivity in the context of 40; enhancing resilience to 137; exposure of degraded farmlands to 62; factoring in targets for meeting domestic food security goals 73; impacts 2, 13, 22, 144, 149; land use changes due to 137–8; mitigation, contributions to 12, 135; modalities of agricultural management in 148; National Adaptation Strategy and Plan of Action for 38; resilience to 144; results in loss of GDP 22; transformational adaptation to 149; UNFCCC 7, 38, 145, 150; vulnerability to 16

climate risks: increased vulnerability to 149; seeking to minimize 38

climate-smart: commitment of stakeholder to meeting responsibilities will increase impact 38; no direct strategy in Nigeria 38; offer potential reduction of yield gaps 143

climate-smart agriculture 6, 16, 22–3, 134; advocated for/implemented through multiple stakeholder groups 137

climate-smart practices 6, 9, 35; Benue region 25, 27; orchards and zero tillage 136; supported by Agricultural Promotion Policy of Nigeria 39

climate-smart village 6, 23; programme 23

co-benefits 10; adaptation and mitigation 11, 135; of intercropping 127; of multifunctional land uses 129; social, economic and environmental 124

Comprehensive Africa Agriculture Development Programme (CAADP) 136–7, 150

Consultative Group on International Agricultural Research (CGIAR) 23; Program on Climate Change, Agriculture and Food Security (CCAFS) 23

Convention on Biological Diversity (CBD) 7

crop production advice and training on 29; discussion with households for information on 55; farmers reclaim all their land for 55; farm visits advising on 37; improved for food security 96; increased under pruned trees 48; intercrop 12; multifunctional parkland with 4; waste water from aquaculture recycled to 84

cultivation access to land for 128; crop, new cycle of 48; declining soil fertility during 3; farmers outside the community may rent land for 25; increased, of natural lands 127; maize 115; rainfed 24, 40; rice-fish 4; rotational/continuous 65; shifting 3

deforestation 14; agriculture-driven 144; agroforestry avoids 11; driven by renting out of agricultural land 8; leakage issues of agricultural expansion leading to 135; Reducing Emissions from Deforestation and Forest Degradation (REDD+) 7; reducing greenhouse gas emissions by halting 38

Index

degradation: environmental 8, 144; forest/forestry 7, 10–11, 135; land 23, 96, **100**, 107, 111, 144; natural-resource 96; soil 148
Department of Fisheries 86
desert date trees 47, 49
desertification 7, 23, 48; United Nations Convention to Combat (UNCCD) 101
drawbacks 9, 148; in developing multifunctional land-use systems 10; of extensive aquaculture systems 81; of smallholder maize-based multiplecropping systems **125**; of total pruning of trees for rejuvenation 48
drivers: for adopting more permanent multiple-cropping systems 66; for adoption of different cassava-based multiple-cropping practices 63; in aquaculture 91; of economic development 109; need to develop land-efficient technologies 119; for peri-urban farming 65–6, 70; of traditional practices 5
drivers towards multifunctional land uses 2; increased demand for food 16; policy 67
drought 31, 100; resistant crops 33–5
drought-tolerant: higher-yielding varieties developed 124; varieties of cowpea 34

Economics of Ecosystems and Biodiversity AgriFood initiative 144
ecosystem services 7, 22; accrued from land use 84; Common International Classification of 147–8; increased production and productivity of 48; provided by some tree species 47; replacing external inputs with 145, **146**; schemes with payments for 12, 143, 145
energy: department 10; high costs for intensive systems of aquaculture 81; renewable 145; required for processing shea nuts 52; solar 89
erosion: control 1–2; control measures 127; ongoing processes 98; prevent 33; status **100**; *see also* soil erosion
Esoko 141
Ethiopia 111; climate-smart agriculture 137; domestic demand fuelled by population increase 121; food-for-work programmes 15; government subsidies discontinued 80; high Gini coefficient value 15; industrialization strategy 110; integrated watershed management 10, 15–16, 107; investment costs for water harvesting 143; maize yields *116*, 119, *120*; policy-driven land use 7; soil and water conservation measures implemented 101; watershed management programmes 142; World Food Programme in 110
Ethiopian government 96–7, 109; Highlands Reclamation Study 96; rainfed maize 119
Ethiopia, northern 98; rural households depend on rainfed agriculture 96
experiment 139; with maize in rotations/intercropping 124
experimental fields 34; research 10; yields 11
experimentation, land users' 5
extension 6, 10; systems 3, 9, 126
extension service (s) 52, 83; community-based 141; gaps in 10; government needs to reform/expand 92; insufficient 79; investing in 129; lack of 74; system (Agricultural Development Programme) 29, 37; target of increased investment in 37; underfunded 141; workers 37
extension service (s), agricultural 25, 71; fading 37; inadequate public 141
extension service (s), access to 71, 82; hindered by poor transport/infrastructure 115; limited 90

Faidherbia albida 14, 47
fallow **65**; fields under 65; land 29, 48, 50, 55; periods 3, 27, 48, 53, 127
family: business of fish farming 86; and community decision-making and management 107; exclusion of women from income-generating activities affects 136; labour 30; labour, cost for 64; land ownership 29, **32**; land and practices having stayed in 30; *Loranthaceae* 55; members, allotting plots to 126; members lay claim on land 128; ventures, small 29
fanya-juu 101, 108
farmers 10, 15, **146**; advantage in savanna region 117; assisted natural

regeneration 54; Benue 25, 34–5; biometric registration of 39; bringing more control and ownership to 140; conflicts with herders 39; continuous improvement of crop varieties 35; convert natural woodland to farmland 47; dependence on subsistence agriculture reduced 111; difficult to obtain land titles for agricultural production 37; economic investment capacity 139; ending involvement in the programme 109; entrepreneurial actors 148; environmental services achieved by 145; expenses high for 36; extension services assist in management of shea parkland 52–3; extension workers for 37; farming systems 65; few have vehicles 70; fish farming 80; food-secure 15; honey bee production 106; identified problem/solution 137; integrate cereal crops with legume 50; investing their gains 74; issue of pruning trees 48; keeping livestock 50; kept indigenous trees 27; knowledge system tools 150; land acquisition process for 38; left agriculture 64; lose confidence in sustainability and effectiveness of new technologies 109; maize growing, monoculture or intercropped 121; optimum value for produce 36; organized cooperative societies/groups 118; participatory action research 23; peri-urban cassava 63; prefer herbicides 33; reclaim all land for crop production 55; reducing negative environmental effects 1; reluctance to test new agriculture solutions 126; reduce tree density by non-replacement 56; registered in Port Harcourt 64; reluctant to participate in the interventions 108; rural 36, 118; with shea trees 27; sustainable trajectories, long-term 148; trade-off calculations 9; transfer of technologies to 108; turned scarcities into resources 134; women 126; zero tillage 32, 33, 40; *see also* peri-urban farmers, smallholder farmers, small-scale farmers

farmers' access: credit and insurance 141; information 141; longer-term credit demanded 141; quality fertilizer at lower cost, improved 119; training, public investment in 128

farmers, assistance for: government scheme reached 36; grants from International Fund for Agriculture and Development and World Bank 35; subsidies on inputs for 15–16; support in adopting beneficial practices 129; technical and control over assets 149

farmers associations/groups 140; commodity cooperative 35; investment among 73; inclusive stakeholder consultation process for 39

farmers, benefits for: agroforestry, multiple 3; multifunctional systems 10; short-term, preferences for 109

farmers, challenges and risks: changing rainfall patterns 40; live with 139; overcoming 9; taken and failed 137; top 31

farmers, credit for 141; Agricultural Credit Guarantee Schemes 128; initiative 7; land cannot be used as collateral 37; loans provided 108; provision of fertilizer 127

farmers, incentivized: adopt sustainable practices 127; cultivate maize lands 124; invest in conservation agriculture 144; not motivated to change practices 135

farmers' income 15; and food provided by trees 40; higher value crops to improve 71; raised 106; revenues increased by good management of trees on parklands 56; richer 16; wealthier 126

farmers, information technology for: computer programme Fertilizer Optimization Tool 71–2; connected to internet 15; SMS service to cell phones 35

farmers' labour: contributions of labour-for-food 142; inputs reduced 4

farmers, new crop varieties for: needs met by 35; registered and released to 34

farmers, orchard 25, 29, 31; fruit-growing 104; and zero tillage 25, **26**

farmers, poor 128; indigent 25; locked into monoculture systems 14; poverty reduction among 119; resource-poor 109 farmers, training for 71, 110; and advisory services 74; awareness-raising activities for 36–7; centres 97; needed 32; public investment in 128

farmers, water availability: enabled increase of farmland area 104; cannot afford long distance transport 56

farmlands: in Burkina Faso 48; distance from markets 70; exposure to climate change 62; feeding growing population on less 143; fragmentation and expansion of 127; increased by farmers 104; keeping scattered trees in 25, 27; land certificates awarded for 108; land-tenure certification procedures for 110–11; mixed maize-based system on 122; more intensive production forced on 64; natural woodland converted to 47; return-on-investment optimized on 124; practices that generate multifunctional benefits from maize 129; prevalent practice of dividing 126; retention of indigenous tree species on 40; soil nutrient balance in 126; traditional, until crude oil discovered 64; used as pasture during dry season 50

female: extension workers 37; headed households 55; tilapia, adult 86

fertiliser: biological nutrient **123**; inorganic, for compost **146**

firewood/fuelwood 1–3, 27, 47, 50, 70, 109

fish 24, 81; annual supply in Kenya 79; appropriate species 134; consumption per-capita, global 78; demand high 78, 89; disease outbreaks 90; escape 90; Farmed Fish Marketing Information System 83; fish-rearing technology 86; health management 91; hormone-treatment technique 88; income from sales 83; local 89; markets 83; market price of 89; most common farmed species 80; movement, restriction on 91; nitrogenous waste removed by bio-filter 86; non-uniform sizes 86; producing quality 85; raising where there is no water 134; rising market prices for 81–2; safe, certification of farm operations for 91; sold at markets 89; stocks 78, 90; trade 82; waste, nutrients from 84; wild, effects of estrogens on 88

fish culture: integrated with horticulture, livestock and forestry 88; systems, intensive 81; tanks, outdoor 87

fisheries 29, 37; collaboration with the Ministry of Agriculture, Livestock and 86; ecosystem approach to 92; employment in 78, 82; guidelines for development of 91; partnership with Department of 86; Research Institute, Kenya Marine and 80; women employed in 83

fish farmers 82; cooperatives 91; farmer-to-farmer networks 90; feed fish on locally available ingredients 82; groups offer training and sharing information 83; increasing needs and demands from 92; integrated aquaculture-horticulture 90; Kenyan 85; learning centre for 86; training for 90

fish farming 6, 15–16, 84, 143; centre of excellence 86; drivers 91; environmental impacts reduced 90; as family business 86; integrated with crops and/or livestock 80; in Kenya 4, 8; organizations for sharing information on 92; pioneered by privately-owned farm 86; prosperous option for small-scale farmers 10; research activities to enhance 90; rural 79; in school curricula to create awareness 92; in semi-arid environments 134; small-scale, in sub Saharan countries 79; struggles and risks common to business development 139; techniques, farmer-to-farmer networks transfer knowledge on 90; training programmes and capacity development in 92; transition from semi- to more intensive 82; women organized into groups to be stronger in market negotiations 136

fish farming technologies: new 79; research is needed to develop 92

fish feed 4; annual demand for 82; imported ingredients for 91; industry, lack of standardized guidelines for 82; meal from dry fresh water shrimp and *Rastrineobola argentea* 82;

Index   161

natural productivity enhanced 81; on organic matter 81; quality imported 86; sourcing of 90
fish ponds: adverse effects of polluted water 90; bottoms covered with plastic liners 88; land allocation programmes for 7; near homesteads 4; on-site reservoir for 86; open, with water storage tanks 85; regarded as household assets 83
fish production 81, 88; from aquaculture 78, 80; annual, doubled 88–9; intensive, most recent system for 86; investment in a borehole increased 86; on Kamuthanga Farm 85; losses due to substandard supplies of feed 82; models for integrating with other farming activities 78; at Musuu Farm 89; national average 89; recirculating systems and constant water temperatures can increase 89; semi-intensive and intensive systems 84; units, re-use of water from 86; water pollution with agrochemicals from horticulture could adversely affect 90
fish supply: annual 79; and demand 89; gap in 86
fodder 1, 3; additional benefits from 106; animal, bought from increased income 70; green, provided by stumped trees 27; growing periods 22; herders expected to cut grass or buy 39; palatable qualities 23; production increased 106; species, tree and shrub 23, 47
fodder from crop residues 124; parkland systems produce 134
food price (s): global 143; global crisis in 16; inflation 15, 143; regional, impact of unstable production on 121; rising 65; subsidies on 15
food security 2, 17, **146**; CGIAR Program 23; challenges of ensuring 22; climate impact studies 11; commercial farms expected to contribute to 143; household 7, 63, 84, 126; improvement of crop varieties to meet needs 35; improved, contributions to 16; Maputo Declaration on Agriculture and 136; managing income disparities required to ensure 15; not solved by increasing crop yields alone 147; parkland agroforestry systems used by farmers for 47; research on 142; role of science in promoting 145; Sustainable Development Goals on 23; sustainable increase of agricultural productivity to enhance 38; of urban low-income net food buyers 143
food security, global: contributions to 11; requires large-scale monoculture production of staple crops 150
food security, national 150; land-use contribution to **146**; maize-based multiple-cropping systems to 128; policies 129; targeted public investment in agricultural sector needed to ensure 36; sustainable agricultural productivity for 39
forest: authorities 56; central zones 117; degradation 7, 11, 135; humid zones 115; institutional bodies 10; legislation 56; mimicking natural 3; mixed 7; natural, severe degradation 11; products, non-wood 53; rain, Nigerian belt 66; reserves, fringes of 127; resources 121
forest management: carbon stock improvements through 53; sustainable 23; tree, department for 56
forestry 7, 11, 64; degradation 10; initiatives, local community 89; integrated crop-livestock-forestry systems 7; integration with 88; sector, greenhouse gas emissions from 22, 135, 141
fruit: continuous cropping 66; diverse sources of 47; in diversified cassava-based systems 71, 72; harvesting and processing technologies for 36; low-hanging 137; oil palm 66; orchards 40, 140; perennial **63**; in peri-urban multiple-cropping systems **69**; plants, selection of 134; pulp 27; range grown on valley plains 104; SAFRUIT project 53; yields, lost 48
fruit production 134; changes in **100**; increased 55; low 55; reduced 48
fruit trees 29, 53, **67**, 71, 124; intercropping systems with 124; partial pruning to improve health 53–4; shea 50, 53

gender 73, **99**; bias 14; of citrus orchard and zero tillage farmers **26**; differences 74; discriminations, traditional 139; division of labour 83; equality 38; equity, improving 83; farm size and cassava yield by 72; gap 71; of household head **69**; imbalances in terms of land tenure 126; inequalities **146**; neutral practices 140; relations 82; role in Kenyan aquaculture 82; roles 82, 140; targeted efforts 83
gendered traditions 15
genetic modification 14; Resources and Biotechnology, National Centre for 34–5; vigour in wild fish stocks 90
Gini coefficient values 15, 61
Global Environment Facility (GEF) 7
grassland (s) 24; conversions 14, 144; expansion of 101
greenhouse gas (GHG) emissions: from agriculture and forestry sectors, mitigating 22; contributions of poor smallholder farmers to 135; global, reducing 23; global, halving by digital solutions 141; inventories 135; from land uses, reducing 10; reducing 7, 38, 89, **146**
Green Revolution 6, 118

high-tech: infrastructure 14; investments in 139; system 139
high-yield 70; varieties 62
high-yielding varieties: cassava 74; choosing 62; maize 124; new crop 34; of white maize 121
homegardens 3, 5, 11; important source for local biodiversity conservation 11; land allocation programmes for 7; multifunctional land use 8; smaller gardens/more biomass 11
homesteads fish ponds near 4; fruit trees in 29; vegetable crops on plots around 50; walking distance to water points reduced 104

incentives 1; for adopting new technologies 149; to adopt sustainable practices 127; cash and grain 109; dependency 108; extension support and higher credit lines 128; for farmers to invest in conservation agriculture and agrobiodiversity 144; financial 92; for making intensive aquaculture more commercially viable 91; to reduce the threat of externalities 91; for value-added processing of farm outputs 128
income 2, **99**, **146–7**; benefits to 83; benefits from maize-based multiple-cropping system **123**; biases 14; from cassava yield 65; contribute to children's education 89; decline 135; diversified 124; from fish sales 83; from fruit during hungry months 28; gross 30, *31*, *33*, *34*; inequalities 15, 61; intercropping/continuous cropping systems contribute to 66; from livestock **100**; losses for crops that perish 36; lost 109; low net food buyers, urban 143; male relatives have interfered to get access to 83; mixed fruit orchards and retention of indigenous tree species on farmlands provide 40; opportunities for women 52, 82; from other crops in orchard 30–1; raised by women-headed households with irrigation 106; role of peri-urban agriculture as buffer of 136; for rural women, increase in 55; from sales per hectare 33; scarcity of 3; from shea butter 48; strengthening 25; from vegetable and spice production 106
income, additional/extra generated from multifunctional land 136; off-farm jobs for cash 9; opportunities to generate 12; used to cover school/medical expenses 106
income, farm 32, 71; increases in 38; profitability ratio per hectare correlated with 31
income generating activities 63, 111; few 52; women's exclusion from 136
income, higher 70; low-cost changes can lead to 35; and more stable 36
income, household: aquaculture contributes to 78; changes in **100**; generating activities, emphasis on 97; doubled by fodder, roof grass and bee keeping 106; increase by diversifying activities 97; poor 52, 142; reliance on household members for labour impacts on 126; share spent on food 15
income improved/increased 106; agricultural inputs contributed to

106; allowed women's participation in community issues 107; by cassava-based multiple-cropping systems 73; for farmers by higher yield or higher-value crops to 71; women's 56
income, low 16; consumers 15; urban net food buyers 143
income source: shea, for rural women 52; for women 50; women pushed out of traditional 15
information technology 71; mobile 83; training in 71
infrastructure 31; agriculture requires investments in 149; green 145; high-tech 14; marketing 78; poor access to 124; poor, hinders marketing opportunities 115; previously established 118; processing, investing in 62; rural 37, 73; social 97; solutions, community-based approaches 109; transportation 90; for water harvesting 96
innovation (s) 5; approached by farmers in different ways 140; dependent on risk-takers 6; farming 6; for improved production 2; Strategy for Africa 2024 150; technological 6
innovative: improvement to challenges Africa is facing 16; practices 6; shea tree techniques, research project on 53
INNOVKAR 53
inorganic fertilisers 35, 64, 96, 101, 106, **146**
Integrated Food Security Phase Classification 117
integrated watershed management 98, **100**, 102, *103*, 104, 106, 110–11, 134, 137, 143; in Ethiopia 10, 15; interventions 100–1, *107*; practices 96, 101, 108, 135; technologies 97, 109
intensification 144; agricultural 10, 143; agroecological 53; of food production systems 92; integrated aquaculture 81
intensive aquaculture/fish farming 82, 91; intercropping **123**; production 64, 83; *see also* semi-intensive fish farming
intensive systems 80–1; aquaculture production 81; of crop rotation **123**; fish culture/production 81, 84, 86; marginalised women in trade and processing industry 82; multiple-cropping 66, **125**; production, commercial 81; production range 86; *see also* semi-intensive systems
intercropping 25, 65; in Benue region 27; cassava 63, 66; co-benefits of 127; farmers can optimize allocation of resources 124; higher LER 66; improved productivity 23, 35; incompatible with machinery for operations 125–6; with legumes 28–9, 66, 140; maize-based 122, **123**, 124; parklands systems 14; practices maximize returns 124; produces more than monocropping 12; recent statistics unavailable 121; systems 12, 66, 124; to utilize biological nitrogen fixation 123; widespread practices 122
internet, connected to 15
investment (s) 6, 39; additional, for agricultural technologies 109; in agriculture and rural development neglected 61; aquaculture 92; avoided crop losses 36; capacity 16, 81; capital, local and foreign 91; costs 82, 108, 111, **125**, 143, 149; demanding modernized solutions 16; economic capacity 139; in extension services 37; in extension systems 40; external financial 111; in farmers' access to training and materials 128; among farmer groups 73; government-led 137; at household level 111; in infrastructure 149; land-use changes require 137; larger, diversification safety net 140; levels for achieving sustainable food security and livelihoods 92; long-term 25, 37, 109, 136, 139; long-term agricultural 38; mechanisms to attract 137; to minimize water use 90; in natural resources management 111; options 135; plans, national agricultural 136; private co-investment 140; private sector 91, 127; public 9, 37, 128; return-on- 72, 119, 124, **125**, **147**; targeted public 36; in trained extension and advisory services 141; in training programmes 92; for water harvesting bunds 143; women have less capital for 70

164   *Index*

irrigation 104, 149; canal 104; changes in **100**, 104; drip *85*, 88; facilities, benefits for women and children 106; improve 98; literature **99**; more water available for 104; from open fish pond with water storage tanks *85*; raised local awareness and knowledge about techniques 111; requires transporting water long distances 56; shallow ponds for 101; small-scale 100; surface, intensified use of 121; water supplies for 97

Kamuthanga Fish Farm 84, 86, *87*, 88–9; centre of excellence in breeding 90; land-use changes at 85

Kenya 15–16; annual fish supply in 79; changes in land use due to climate-change 137–8; climate-smart agriculture 137; Economic Stimulus Program 80; fish farming in 4, 8, 16, 79; gaps in extension service 10; integrated aquaculture systems in 79; maintenance practice in orchards 29; maize harvesting *120*; marketing challenges for women 83; *mpesa* 141; off-farm agriculture work 15; patriarchal society 82

Kenya, aquaculture in *80*; land use studies 84; production in 81; gender division of labour 83; Kamuthanga Fish Farm operation 89; sustainable development 81, 91

Kenya Marine and Fisheries Research Institute (KMFRI) 80

Kenyan aquaculture 79, 82–3; drylands, studies from 109; farmers sought alternative farming activities 85; government's economic agenda 90; government's Eat More Fish campaign 79

labour 14, **32**; available, women have less 70; cheap, availability of 127; communities contributed free days 101; costs 25, 30, 32–3, 64, **146**; dependent **147**; family 30, 64; farmer 4; farmers can optimize allocation of resources 124; -for-food, contributed by farmers 142; gender division of 83; hired 12, 31, 64; household 31; intensive **125**; involved in processing butter from nuts 52; reduced in maize-based multiple-cropping systems 124; rural 15; scarcity of 2; seasonal 9, 12; shortage 31, 33; skilled 54; source of **26**; time, inefficient 136

labour inputs: costs for 30; daily, dependence on 9; of food-insecure rural households 108; for maintenance, reduced 4; reduced on maize-based multiple-cropping systems 124; trade-offs on 12

Land Equivalent Ratio (LER) 12, *13*, 14, 25, 29, 63, 66, 70, 79

land fragmentation 14, **125**, 128

land fragmented 16, 71, 119

landscapes 2, **146**; adding trees in 10; beautification 68; climate properties of 13; fewer trees remained in 48; green 66, 101; groundwater tables and biomass brought back to 135; holistic view of 135; increasing tree cover 88, 101; integrated management 22; integrated watershed management practice 134–5; multi-functional 2, 5–6, 10, 47; practice to manage water resources 134; re-greening 144; restoration 40, 101; scale 8; semi-arid 135; transformed by establishment of enclosures 138

landscapes, rural: Burkina Faso 48; Sub-Saharan African 6

Land Use, Land-Use Change and Forestry sector 135

large/larger-scale 38, 48, 92, 121, 139, 142–3, 144, 150; adoption of policies 139; agricultural investments 38; land acquisition 143–4; monoculture production of staple crops 150; producers 121

large-scale interventions: in extension 142; water management 139

legumes 12, 34, 71, **125**; in agroforestry parklands 47; improve nitrogen fixation 29, 123, **125**; integrated with cereal crops 50; intercropping with 10, 28–9, 66, 117, 122, 124, 127, 140; micro-nutrient uptake of maize after **123**; providing input subsidies on 117

livelihoods 2; from agricultural activities 98; aquaculture-based land use affects 84; changes in 104; development and improving 110; earning from staple food crops 137;

equality benefits 136; improving 78; influenced by gender relations 82; investment level for achieving 92; of local communities, contributions to 84; low investment cost to establish and generate **125**; multifunctional land uses contributed to improve food security 16; options provided 9; positive impacts by integrated watershed management interventions 101; restricted by water deficiency 97; sensitive to variations in rainfall patterns and long-term warming 22; shift from previously nomadic/semi-nomadic 7; sources disrupted 143; stress 84; in Sub-Saharan Africa depend on maize cultivation 115; sustainable 97, 110; VI-agroforestry focuses on improving 6

livelihoods, rural 2; benefits brought by multifunctional land uses 128; Kenyan patriarchal society influences 82; in northern Ethiopia depend on rainfed agriculture 96; parkland agroforestry systems essential for people in semi-arid West Africa 56; role of agroforestry parklands in 47

livelihoods, smallholders' 142; loss of multifunctional 6; peri-urban 63

livelihoods, women's 16; changed opportunities for 82; homegardens related to 11; importance of shea for 52, 55; opportunities in aquaculture limited 83

livestock 50, 85–6; agencies for 91; based agropastoral systems 109; crop residues as fodder for 124; diverse species used **146**; extension agents advise on 37; farming integrated with crop/forestry systems 7; fodder qualities to support 23; improved agricultural systems and practices for 38; integrated farming with 65; integrated fish farming with crops and 80; interference, exclosures to keep land free from 96; kept near homes 4; land allocation programmes 7; loans to farmers for the purchase of 108; messages on iCow 141; on Musuu Farm 88; open grazing prohibited 39; producers 90; rotation system with **123**; separated from cultivated land 3; systems in Tanzania 145; water supplies for 97

livestock production 29, **65**; changes in 100; combined with field operations 117

loans 7, 90; accessed through commodity cooperative groups 35; Central Bank of Nigeria 35; from development banks 150; guarantees 91; interest on 65; at low interest rates, women's banking group members 89; private sector Anchor Borrowers Scheme provided 35; return on **147**; services, expansion of 108

loans to farmers 108; additional incomes used for paying back 136; land cannot be used as collateral 37; low-interest 7

maize 8, 11, 62, **125**; area *118*, 119, *120*; area cultivated declined 126; area in Egypt 121; area harvested 118, *119*; in diversified systems 135; Ethiopian 119, 121; farmlands 129; field *28*; imports/importer 117–18; input subsidies on crops 117; integrated in agroforestry parklands 47, *49*; leakage issues of agricultural expansion 135; LER with cassava 70; livelihoods from annual crops of 98; matures in succession 122; Nigerian 121–2; output 115; population relies on 64; producers *120*, 129; rainfed high-yielding varieties of 121; in rotations 124; southern Nigerian region 122; supplemental fish feed 81–2; in traditional farming systems 84; zero tillage common for 32

maize-based systems 9, 115; co-benefits of intercropping in 127; popularity and fast expansion of 122; research conducted on 122, 137; smallholder, potential negative impacts of 125; those with legumes or cereal most widely practised 127

maize cultivation 115; for household consumption/national grain production 124; with legumes **123**; in traditional ways by smallholder farmers 124

maize intercropping 29, 121, 124; with cassava 66, **67**, *68*, 72; co-benefits of 127; to utilize biological nitrogen fixation 123

166　Index

maize multiple-cropping system 63, **69**, 119, 121; in Nigeria 115, *116*, 117, **123**, 127–8

maize production 119; African 115, *120*, 123; benefitted by National Special Programme on Food Security 119; decreased 121; increased 106, 118; integrated 6; intercropped with other seasonal crops 117; Nigerian 117, *118*; Nigerian savanna region 122; policy interventions to improve Nigerian 129; remained relatively unchanged 126; stagnating trends for 117

maize varieties: high-yielding 124; improved 29

maize yields *14*, 129; average 105, *116*; average, Nigerian 119; higher than monocultures 29; increased 105, 126; increases when planted with nitrogen-fixing crops 123; micro-nutrient uptake after legume **123**; potential 119; of smallholder farmers 115

Malabo Declaration 72, 136–7, 150

malnutrition 22, 78, 89

Managing Environmental Resources to Enable Transitions towards more Sustainable Livelihoods 97, 110

manure 50, 64; green 68, 73; management 7; organic 23; recycled for compost 4

Maputo Declaration on Agriculture and Food Security 136

market changes 70; adaptation to 136

market information: accessing 71, 141; gender-targeted efforts linking women to 83; System, Farmed Fish 83

marketing 8, 73; challenges for women 83; Information System, Farmed Fish 83; infrastructure, poor 78; opportunities hindered by poor transport and infrastructure 115; strategies in fish-farming 86

market-oriented 148; primary objective 1

market price for fish: average 89; rising 81–2

markets: access 83, 110, 128, 144; commodity outputs 13; create 1; demands 1–2, 7, 90; deregulation of input 127; distance to/from 70, 83, 136; diversity of channels 149; expansion 90; failures 117; fish 83, 89; governments interfere in 16; growing 50; how to enter 148; international 52, 56, 91; investment required in 149; led agricultural transformation 73; local small-scale 115; negotiations 136; price of crop 72; related knowledge 148; requirements, respond to 52; review of 135; road networks to reach 31; role in driving multifunctional practices **147**; segments 7; saturation, avoidance of **147**; supermarkets 15

market-value chains 136, 140

microclimate: regulation 1–2, 10; with multiplecropping systems in Nigeria **125**; by reallocating water 134; by shade tree 8; by shea parklands 50

millet 11; agroecological intensification of 53; in agroforestry parklands 47, 49; intercropped 27, 121, **123**; rainfed cultivation of 24; yield *14*, 48, 53; zero tillage 32

Ministry of Agriculture, Livestock and Fisheries 86

mitigation: climate 2, 6–7; climate change 12, 39, 135; co-benefits of agroforestry 11; in-situ benefits/losses 135; interventions, profitable 145; winwin interactions with adaptation 135

mobile phones 39; services for agriculture 141; technology to connect to credit 83

monoculture 1, 9, 136; in Africa 63; bias towards 10; cassava 63, 66, 68, 70, 73; contribution to national food security 128; Green Revolution's promotion of 6; Land Equivalent Coefficient 27; large-scale producers practicing 121; maize producers 129; second maize crop 66; systems 12, 14, 124

monoculture production 12, 28, 150; of staple crops, large-scale 135, 150

monoculture yields 11, 29, 35; loss 68; of maize **123**

multifunctional land use 1, 6–7, 129, **146–7**; additional incomes generated from 136; aims to produce more than one product or service 134; assessing 11; carbon pool of 10; case studies across Africa 16; certification schemes involving 8; development of new 16; drivers of 2, 6; enhanced 150; global extent of 8–9; hampered by broad and undefined scope 10; lack of research around 8; measure of 12; multiple functions of 145; opportunities for 16; peri-urban 5; research on 150; resulting from unplanned responses to changed conditions 137; survival without subsidies 142; sustainable 13; transitions to 136

multifunctional land use, benefits of 137; for Africa 142; brought by smallholder farmers 128; capitalizing on 138

multiple-cropping practices: cassava-based 63; maize-based 124

multiple-cropping systems 63, 68, 129; benefit of 66; benefits rarely recognized 127; cassava-based 63–4, 70, 73–4; driver for adoption of 66; farm gains from 64; land use by gender 72; in Nigeria **123**; in Nigeria, none directly promoted 127; peri-urban **63**, **65**, **68**, **69**, 72–3; small-scale 125; supported subsistence and household food security 126

multiple-cropping systems, maize-based 115, 117, 119, 121, 124; comparative advantage of 128; contribution of Nigerian smallholder 128

Musuu Farm 85; awarded the Head of State Commendation 89; contributed to reducing hunger and malnutrition 89; enterprise centred on contributing to needs of local communities 92; integrates fish culture with horticulture, livestock and forestry 88; struggled with low production 90

National Adaptation Strategy and Plan of Action for Climate Change 38

National Centre for Genetic Resources and Biotechnology 34–5

Nationally Determined Contributions (NDC) 7, 11, 137, 150

National Special Programme on Food Security 119

natural disasters 10, 139

néré (Parkia biglobosa (Jacq.) R. Br. ex G. Don) 27, 47–9

Nigeria 3, 6, 8; agricultural extension service system 37; Benue State climate-smart agriculture practices in 27; discontinued government subsidies 80; expected to double its population 61; fertilizer use 121; fruit orchards 140; government interference in agriculture, markets and trade situations 16; intercrops of roots and tubers 127; Land Use Act 38; launched Agricultural Promotion Policy 73; loans from Central Bank of 35; net food importer 61; no direct strategy for climate-smart agriculture 38; peri-urban agriculture case in 15; research on improved crop varieties 23; rural resource centres established 141; semi-arid 23; signed Sustainable Development Goals on poverty and hunger reduction 72; staple foods 61; tested computer programme Fertilizer Optimization Tool 71–2; training programmes for urban and peri-urban farmers 73

Nigeria cassava production: cassava-based multiple-cropping system in 63–4; levels 62; planting periods 66

Nigeria maize production *118*, 119, *120*; integrated 6; intercropping with 121; interventions to improve 129; maize-based systems *122*, **123**, 127; major expansion of 118; net importer 117; output 115, *116*; smallholder multiple cropping systems **125**

Nigerian: Agricultural Land Development Authority 118; children, malnourished 22; croplands, soil conditions in 124; maize 119; Naira 65; National Policy on the Environment 73; people 61; rain forest belt yields of cassava 66; savanna region 121

nomadic: pastoralists in Kenya more sedentary over past decades 138; and semi-nomadic livelihoods 7

nutrition 2, 11, 89–90, 149; improved 23, 111, 128; security 83, **100**

nutritional: diets, enhanced 143; diversity 111; intake **125**; value 11, 50

nuts provided by African agroforestry parklands 47, 134; *see also* shea nuts

off-farm: agriculture work 15; jobs for additional income 9

orchard farmers, citrus 25, **26**, 27, 29; honey production 31 orchards: citrus 25, **26**, 27, 29, *30–1*, 35; climate-smart agriculture practice 136; cost for labour and inputs 33; farms 30; fruit, in Nigeria 140; growing 25; maintenance in 29; mixed fruit 40; more input-demanding than zero tillage 31; other crops in 31; profitability **32**; small family ventures 29; time to establish 30; women granted intercrops in 37

organic 144; fertilizers 35, 81, 96, 101, 106, 127; inputs, use of **146**; manure or compost 23; nutrients, recycling 29; residuals, burning of 29; vegetables 8

organic matter: fish feed on 81; retention of 29; soil enhanced 32; soil increasing *103*, **125**, 140

Organisation for Economic Co-operation and Development (OECD) 1–2

ownership 10, 25, **32**, 56, 64, 140; communal/individual 128; of private ponds 83; sense of 110; sovereign 128; state 108, 111

parasite (s) *Tapinanthus spp* 48, 53, 55; weed (*Striga gesnerioides*) 33–4

parkland agroforestry 47, 56; keeping/managing shea trees on 55; land pressure reduces area of 56; legislation on 56; management 53, 56; pruned and unpruned shea trees on 54; reduction in tree density on 55; trees fall under forest legislation 56

parklands 8; actions to restore 48; dominant trees in 47; gendered traditions turned into opportunities 15; good management of trees enhances food security 56; groundwater recharge 51; herders use for pasture in dry season 50; multifunctional 4; presence of shrubs and trees improved soil carbon and crops 47; soils, manure used to fertilize 50; soil-water interaction demonstrated on shea 134–5; sub-humid zone of West Africa 47; traditional 27; traditional collector access to shea nuts and trees 52; tree-ageing and parasitic infestation 53

parkland shea systems 49–50, 144; with intercropping 49; keeping/managing trees 55; management 53, 56; pruned and unpruned trees 54; regulate the climate 50; stakeholders in production 52; traditional collector access 52; tree-crop interactions and trade-offs 57

parkland systems: intercropping 14; production, poor performance of 48; with scattered trees, functions of 134; traditional multifunctional land-use 3; underutilized 9

pastures 7; change in **100**; farmland/fallow land used during dry season 50; for herders' animals 39; herders use parkland during dry season 50; management regimes 23; parklands used in dry season 56; spaces between scattered trees used during dry season 47, 49

payment agroenvironmental 6; for creating markets and demand 1; for ecosystem services 12, 143, 145; farm inputs provided 39; services from Esoko 141

peri-urban agriculture 15, 71, 135–6; agroforestry 7; areas of Port Harcourt 64; land, scarcity of 71; land use in Port Harcourt, changes in 70; multifunctional land 5; multiple-cropping systems **65**, **68**, **69**; needs to be met 74

peri-urban farmers 70–1; advantages for 70; benefit of multiple-cropping systems 66, 68, 72–3; cassava 63; cassava-based systems 66; driver for 65–6; enabled to develop multiple-cropping systems 72–3; household benefits for 70; little access to agricultural extension services 71; meeting needs of 74; smallholders, livelihoods of 63; training programmes for 73

peri-urban farming 8, 66, 73; cassava-based systems **63**, 64, 67, 137; cropping systems 16; multiple-cropping systems **63**, 68

policy/policies 149; on access to markets 128; advocates 83; African smallholder 15; aimed to ensure household food security 7; aquaculture 81; aspects 38, 556, 72, 90, 109, 127; bias towards monocultures 10; craft 91; and credit 35; directives and programmes

110; effort 74; to enable/discourage multifunctional practices **147**; on the environment, Nigerian National 73; in Ethiopia 110; to facilitate decentralized and participatory development 109–10; fail to promote food security 72–3; favouring maize in multiple-cropping systems 117; federal and regional, adherence to 110; fertilizer 126; on food crops 16; food security component of 90; formulation 15; framework, inadequate 82; Future Award 101; government 38, 126–8; green belt 127; Growth Enhancement Scheme 39; guidelines 81; historical development 117; impacts at national/international scale 135; implementation 15; on import regulations 91; informed processes 139; initiatives 128; instruments for 83; instruments, focusing on 73; integrated 16; interventions 128–9; and investment options 135; limited influence on land-use systems 5; literature **99**; major 110; makers 13, 37, 97, **147**; making, supporting 137; mandates for soil fertility management 127; with multifunctional land use 7; national food security 129; needed to reduce cassava food deficit 73; promoted multiple-cropping systems 127; not aligned with national land use act 128; paradigm shifts 16; that reduce land conversions 144; regime, international climate 139; research-informed 138; revised comprehensive 91; revolutionary 117; serve to improve community efforts 97; for specific goals 2; state and national 40; strategy, concerted 91; structured, to guide local planners 127; successful 129; targeted improved access to inputs 118; targets 10, 127; on tenure 139; water-management, large-scale 139; *see also* agricultural policy/policies

policy/policies drivers 6–7; of processes in Africa, none 148

policy/policies support (ing) 16, 38–9, 109, 127; to encourage farming 119; lacking 127; public-private and local-international partnerships 91; shea parkland management 56

policy-driven: land use 7; process 6

pollinators 3

population: agriculture main occupation 22; Benue State 23; Burkina Faso, rural 55; density 98; earn living from agricultural activities 98; farming 39; food-insecure 61; global urban 61; pressure on land 73, 96; relies on tropical crops 64; Rivers State 64; rural, empower to produce adequate food 128; Tigray, rural 98; trends in Nigeria 115, *118*

population, fish: over 86; wild 90

population growth/increase 2, 13–14, 16, 22, 39, 48, 55–6, 64, 117, 121; expected 78; feeding on less farmland 143; global 61; looming 149; urban 66

post-harvest: losses 36, 78; processing 9, 149; production, value added by women 83

Productive Safety Net Programme 97

promote/promotion 6; agricultural productivity 110; agricultural, strategy document for 117; agriculture as driver of economic development 109; agroforestry 38; biometric registration of farmers 39; cropping systems with legumes 127; effective management of natural resources 39; grafting to shorten vegetative phase 53; Kenyan government's 'Eat More Fish Campaign' 79; maize production 119; of monocultures 6; multifunctional practices 137; multiple-cropping systems with maize 127; national agency for non-wood forest products 53; natural-resources management 110; orchards/retention of indigenous tree species on farmlands 40; Policy, Agricultural 39, 73, 119, 127; ranches 39; of single adaptation responses 149; sustainable agriculture 23; sustainable land-use practices, role of science in 145; sustainable water and land resources management 110; traditional weeding methods 37; of viable aquaculture investments 92; water harvesting by local communities 89 prune (d)/pruning 25, 27, 29, 48, *54*; and assisted natural regeneration 55; more sunlight penetrated and soil fertility increased under 53; partial or total 48, 53–4; requires skilled labour 54; for tree health or rejuvenation 53, 56

rainfall 102; adaptation responses to 150; failures, frequent 84; increasing variability in 62; limited 92; projected to decrease 22; sparse 98; suitable 117; variability, farmers' response to 36; variable 25, 34–5; variable onset and cessation of rainy seasons 24
rainfall, annual 104; average 98; confined to two months 98; long-term average 84; total 24, 84
rainfall patterns: adjust farming calendars in response to 40; differences in 48; variable, climate-smart practices resilient to 35; variations in 22
rainfed: crop cultivation 40; cultivation of arable crops 24; Ethiopian maize 119; farmlands exposed to climate change 62; fields converted into irrigated land 104; high-yielding varieties of white maize in South Africa 121; Nigerian maize 119; practices dominate in Tigray 100; systems, support for year-round production 117
rainfed agriculture 22; farmers' dependence on low-productivity subsistence 111; plays important role for many farming systems 143; previously 104; risk of crop failures in 22; rural households in N. Ethiopia depend on 96
rainwater 54; harvesting ponds 100; and surface run-off 88
rain water harvesting 23; bunds 143; infrastructure for 96; local communities inspired to promote 89; methods 101, 111; practices 88; small-scale household 100
Reducing Emissions from Deforestation and Forest Degradation (REDD+) 7
reforestation 7, 11; Agriculture Promotion Policy targets 127; initiatives 89
remote sensing 8, 144; tools 144

resilient/resilience to adverse climatic events 48; climate 143; to climate change, enhancing 137; climate, information technology for building 71; economic, improve 12; environmental 6; to environmental degradation and climate change 144; to external stress 145; plans for **147**; systems of farming 71; to variable rainfall 35
restoration: of degraded forests and pastures 7; of degraded land 111; initiatives 102; landscape 40
restore: landscapes 101; parklands, actions taken to 48; soil fertility 4, 48, 134
rice 8; fish cultivation 4; intercropping 122; paddy fields 4; rainfed cultivation of 24; supplemental fish feed made from 81–2; yields 4, *14*; zero tillage common practice 32
risk (s) 10, 12, 139, 145; animal feed sufficiency 22; associated with storage of produce 70; balance with rewards **125**; climate 38, 149; of crop failures, reduce 22, 28; current, responses to 150; of damaging other trees or crops 9; farmers decided to take 137; in fish farming 90; identifying 139; of losing the whole harvest 12; major for farmers 149; natural disaster 139, 145; in persistent use of herbicides 36; potential **147**; of push for cheaper food 15; reduce and spread 107; reduction strategy **125**; substandard supplies of feed posing 82; technologies that can reduce 149; of urban hunger increasing 61; of young plants being grazed by roaming animals 56
risk-takers 6; risk-taking strategies 137
root (s) 47; biomass 124; carbon sequestered in 51; crop 9, 12; decomposition 51; depths, mixed 104, 124; host 33; intercrops of 127; processing and storage 37; stock, rough lemon 29; tree 27, 103; tree, deeper-rooted **63**; tuberous cassava 62; zone, rainfall percolating through 104
rotational cultivation 65; slash-and-burn practice 3

rotations, crop **63**, **65**, 127, 129; maize 121, **123**, 124; managing 124; systems 65; with tubers 32

scale 142; benefits of 9; commercial farms, medium- 143; economies of 115, **125**; impacts at **147**; integrated water management, basin- 138; landscape 8; loss 2; national and international 135; NGO sector achieve 140; scheme for registered farmers 39; temporal and spatial **147**; time, more positive outcomes over longer 149; watershed 96; *see also* large/larger-scale, small-scale
scale-up: enterprise efforts 97; public and private programmes, actions to 92
school 86; children's opportunities to attend 126; curricula 92; extra income used to cover 106; fees 50, 70; Productive Safety Net Programme built 97
science, role of 148; in African agriculture 148; role in promoting sustainable land-use practices 145
Science Technology and Innovation Strategy for Africa 2024 150
semi-arid: Africa, restoration of degraded land in 111; Benue State of Nigeria 23; climate 84, 98; environments, fish-farming in 134; landscapes, groundwater tables and biomass brought back 135; Sub-Saharan Africa 107; West Africa, livelihoods of rural people in 56
semi-arid areas/regions 88, 92; benefits of watershed management with tree cover 104; of eastern Kenya 84; trees in 47; integrated watershed management strategy 97
semi-intensive fish farming 82; fish production 89
semi-intensive systems 80–1; fish production 84, 89
shade 48, 55; cassava provides 66; competition for 9; main cause of yield decrease 48; positive effect on millet yields 48; preferred ecosystem function of 68; provided for ponds 88, 134; reduces temperature below canopies 12; trees 8, 48, **63**
shea 141; agroforestry parkland 49; caterpillars 50; collection, processing and trading, government projects on 52; conservation 53; food crops production on agroforestry parkland 50; fruit 50, 53; fruit production improved by pruning 53; importance for rural women's livelihoods 52, 55; industry, sustainable development of 52; leaves 27; parklands 49, 50; processing and business skills 140; products 50, 56; rejuvenation 53
shea butter 50; incomes from 48; international market for 56; processed and traded 51; production 27
shea nuts 3, 50, 52; grafting to improve quality 54; processing 53, 56; processing and trading 53; revenue earned from sales 55; yield 55
shea parklands 49, 144–5; demonstrate soil-water interaction benefits 134–5; expansion/success of management practices 56; management of 53; regulate the microclimate 50; trade-offs in 57
shea production 3, 9, 52; stakeholders in 52
shea trees (*Vitellaria paradoxa* C.F. *Gaertn*) 3, 9, 16, 27–8, 47; affected by African mistletoe species 55; assisted natural regeneration 54; buffer against desertification 48; fruit production 53; important for honey production 50; innovative techniques, research project on 53; keeping and managing on parklands 55; listed as protected species 56; management practices for 53; most common species 47; old, fruit production 53; partial pruning 48; products and functions 50; protection of seedlings during dry season 56; pruned and unpruned 54; sequester carbon in trunk and roots 51; suitability mapping of 144; women harvest 49
shifting cultivation 3, 11
shrubs: annual crops mixed with **63**; degraded grazing lands re-greened by 101; fodder 23; presence improved soil carbon/crop yields 47; scattered, cultivation between 47
smallholder farmers 81; approached innovations in different ways 140; bring multiple benefits to national agenda 128; coping strategies

smallholder farmers *continued*
  adopted by 127; at core of agricultural-development-led industrialization strategy 110; cultivate maize in traditional ways with other crops 124; economic benefits motivating for 12; often forgotten 142; with fragmented lands 18; hatched fish eggs sold to 86; maize cultivation by 115; maize yields 115; in Nigeria grow cassava with mixed farming systems 62; poor 135; poorly equipped to manage soil nutrient balance 126; power imbalances unlikely to benefit 139; relatively flexible 124; source of information given to 148; yields 11
small-scale 110; aquaculture 91; farms 125–6; fish-farming 79, 81; irrigation 100; markets 115; multiple-cropping systems 125; rainwater harvesting ponds 100; resource-poor farmers 109
small-scale farmers 10, 129; contributions to multiple Sustainable Development Goals 129
soil 73, **125**; bunds 101; deeper 98; degradation 148; disturbance reduced 33; dominant reference group 98; hydraulic properties improved 51; improved the infiltration capacity of 103; improvement with organic/inorganic fertilizers 101; information assessed 55; interaction with water 134–5; loss halved 103; management messages on iCow 141; movement of 101; net annual loss from croplands 96; parkland fertilized with manure 50; properties improved 51; quality *103*, 124, **125**; restored by lying fallow 48; retention of water in 29; structure 3, 32, *33*; structure improved 3; structure maintained 32, *33*; tests 37; trampling by cattle compacts 39; zero/minimum tillage 32, 35, 40
soil carbon: improved by trees and shrubs 47; increased 29; restoring 134; retained 40; stocks increased 135
soil conservation 1, 96, 127; maintenance 108; measures 101, 111; technologies 23
soil erosion 12, 32; bamboo trees planted for preventing 85; literature on **99**; ongoing processes 98; preventing *33*, *85*; prevented by planting without tillage 32, *33*; protection from 12; reduced 88, 103, 134; reduced by increased tree/vegetation cover 88, 103; sheet and rill 103; speed of 96; status and change **100**
soil fertility 3; assisted by some tree species 47; compost to restore 4; decline 3, 48, 53; dictates spacing of cassava 66; management 23, 127; restoring 134
foil fertility improvement/increase 96–7, 105, 123; by mulching 29, 48; under pruned trees 53
soil health **99**, 101; improved by nitrogen 29; indicators 103; public goods 13; stability improved by root systems 12, 27; stabilize by agroforestry 3; status improving 10; storage of carbon in 12
soil moisture 12; conserving 32, *33*; holding capacity 9; improved by mixed root depths 104; improved by shade provision 48; increase due to hydraulic lift mechanism 51; maximized by intercropping **63**; retention of 29
soil nutrients **63**; balance 126; crop nutrient uptake enhanced in 140; management 123; poor 62, 124, 134
soil organic matter: improved 32; increased 103, 140; retention of 29
soil, poor **123**; recovering 134
soil regeneration: delayed by permanent cropping 53; natural 96
soil surface: evaporation from 12; plant residuals left to decay on 32
sorghum 11, 98; agroecological intensification of 53; areas converted to maize 121; integration in agroforestry parklands 47; intercropped 121–2, **123**; shea tree intercropped with 27
sorghum yield *14*, 47; decreased in shade 48; increased 105; increased by pruning trees 53
soybean 24, 29; bring increases in maize yields **123**; farmers 35; input subsidies on 117; intercropped *122*; processing 37; stalks, thatching alternatives 36

stakeholders **147**; commit to meeting responsibilities 38; engagement of 119; in fish farm groups 83; groups, multiple 137; inclusive consultation process 39; in shea production 52; in value chain targeted by Blue Growth Initiative 92

staple crops 7; commercial production 66; harvesting and processing technologies 36; importance for food security 115; intercropping with legumes 140; locally processed/stored 115; monoculture production of 150; productivity 14; provide multiple benefits in diversified systems 135

stress-tolerant crops 11; seeds 96, 143; systems 23

technological innovations in multifunctional farming 6

technology/technologies 14, 144; adoption constrained by gender imbalances 126; biotechnology 34–5; difficult to apply to scale **125**; farm **125**; for high production 81; improves 144; National Centre for Genetic Resources and Biotechnology 34–5; new, study on suitability to the area 109; Science, Technology and Innovation Strategy for Africa 2024 150; simple 89; support through Economic Stimulus Program 86, 98; traditional 98; *see also* high-tech, information technology

technology in fish culture systems 81; changes in 15; fish-rearing 86; recirculating aquaculture system 88–9

temperatures 8; cassava yields sensitive to increasing minimum 62; constant 88–9; regulate stomata and photosynthesis functions 12; semi-arid climate 84; shade reduces 12; Tigray average annual 98; trees reduced 48, 50–1

tenure: ambiguous 40; insecure 139; laws related to 38; practices 110; restricted 108; security 108, 144; status insecure 56; system, current 128; systems, weak 39; uncertain 9, 16; voluntary guidelines for 139

tenure, land 14, 37, 52; certification 110–11; conditions required for 70; gender imbalances in 126; insecure 25; policy intervention critical for 128; uncertain 124

three-dimensional farming 6, 11

trade: barriers reduced 1; deficits, agricultural 142; fish 82; intra-African agricultural 137; liberalization policies 16; manufacturing surpluses 142; in shea nuts 52–3; situations vary 16

trade-offs 9; assessments 135; economic 135; on labour inputs 12; in shea parklands systems 57; temporal or spatial 150

traditional: agroforestry systems 135; bee hives 50; collector access to shea nuts and trees 52; consumption of shea caterpillars 50; farmland until crude oil discovered 64; fish trade offered income opportunities for women 82; income sources, women pushed out of 15; menus 66; ownership of land by men 37, 49; parklands 27; processing butter from nuts 52; technology 98; view of farmers as risk averse 10; women's practices 140

traditional farming systems/practices 3, 84; abandoned 84; cassava landraces 66; culture 24–5; drivers of 5; field clearance 32; growing coffee plants 8; knowledge about 110, **146**; low-cost adaptation of 16; planting 53; subsistence 2; weeding methods promoted 37

traditional grain crops 122; maize cultivation 124; varieties replaced 35

traditional land use (s) 3, 137; multifunctional 3, 6

transformation agricultural 38, 119, 150; of agri-food system 143; of conventional farms 8; market-led agricultural 73; rural 16; socioeconomic 149; *see also* Agricultural Transformation Agenda 119

transformational: adaptation to climate change 149; changes in Niger 149

tree-based systems 50; training farmers on 141

tree cover 9; afforestation programme to increase 88; increased in drylands 144; retention of indigenous species promotes 40; watershed management with 104 tree density 27; assisted natural regeneration helps increase 54; declining trend in 48; determined by value of products and services provided 47; reduced by parasitic African mistletoe 55; reduced when trees are unproductive 56; reduction on agroforestry parklands 55

tree planting 6, 23, 53, 104, 127; non-landowners forbidden 49; plantations 25, 145

tree pruning for health or rejuvenation 53, 56; total, fruit production in 54; total, for rejuvenation 48

trees 3, 4, 8; belong to land owner in Burkina Faso 10; crops, perennial 24; crops, transition to 29; deeper-rooted 63; dominant in parklands 47; enclosures made from living 138; greater diversity of 134; integration in agriculture and farming practices 144; interactions with crops 48, 57, 71, 135; long-term management practices 56; mortality 55–6; non-productive, elimination of 55; not replaced 48; owned through patriarchal lines 37; parkland, under forest legislation 56; positive interactions with crops 12; reduce negative weather impacts 10; risk of damaging 9; standard spacing between 126; stands, permanent 139; store more carbon in 12; stumped 27; timber 12, 63; traditional collector access to parklands 52; unauthorized management prohibited 56; *Vitellaria paradoxa* 9

tree seedlings: farmers discouraged from keeping 56; nursery 141; survival rate improved 101

trees, fruit and citrus 29, 31, 53, **67**, 71, 124; intercropping systems with 124, 134; local knowledge about 53; production in total pruning 54; sources of fruits 47

trees, scattered/dispersed 3, 24–5, 47, 134; on grazing lands 134; in savanna areas 124

trees, shade 48; canopies 8, 48, 50; ponds and reduce evaporation 88

trees, soil improved by: accumulated/increased biomass 51, 103, 105; carbon and crop yields in parklands 47; hydraulic properties 51

tree species bamboo 85, 88; fertilizer 12, 14; fodder species 23; indigenous 27, 40; ironwood 28; kept as standing trees 27; non-landowners forbidden to harvest certain rare 49; palm 66; provide important ecosystem services 47; shea most common 47

tubers: cassava 62, 70; crop rotations with 32; intercropping 122, 127; harvesting and processing technologies unavailable 36; high-yielding drought, disease and pest resistant varieties 34; integrated in parklands 47; potato (*Solanum tuberosum*) 105; processing and storage 37

UN Framework Convention on Climate Change (UNFCCC) 7, 38, 145, 150

value chain 9; agricultural 73; of agricultural products 148; market 136, 140; of multifunctional farming practices **146**; post-harvest processing stages of 149; starts with improved stress-tolerant seeds 143; sustainability endangered 81

value chain, fish trade 83; development of 91; input side 82; stakeholders targeted by Blue Growth Initiative 92

vegetables 64, 105, **125**; in cassava multiple-cropping system 63; changes in productivity **100**; high-yielding drought, disease and pest resistant varieties 34; home produced 3; incomes tripled 106; integration with other crops 47; intercropped with maize 117, 122; intercropping with cassava 66, **67**, 68, **69**, 70, 72; irrigated with nutrient-loaded water from aquaculture 88; organic 8; preservation 37; rainfed cultivation of 24; sole crop on smaller homestead plots 50

vertical farming 6, 11
vertisols and calcisols 98
VI-agroforestry 6

water deficit/shortage 89, 135
water infiltration 1, 134; capacity of the soils 103; increased 104
watersheds 96, 98, 101; Abraha We Atsbeha 106; average soil loss halved

103; benefits of investments 111; benefits of re-greening 103; community-based management 96, 110; Enabered 103; *fanya-juu* bunds 108; farmers reluctant to participate in interventions 108; Gulle 101, 103–4, 106; high species diversity 102; Hintalo 101, *102*; honey production increased 106; implementation 106; increased soil fertility and water availability 105; interventions 106; literature **99**; management programmes 96, 111, 142; Mariam-Shewito 104, *105*, *107*; Medego 103; Messebo **99**, **100**, 101, 106; non-rehabilitated **100**; participatory management 110; rehabilitation and ecosystem improvement 7; Sheka **99**, **100**, 104, 106; socioeconomic benefits observed 104; soil quality parameters higher with treatments *103*; successful 109; surface run-off 103; technologies 109; in Tigray **99**; untreated 106; water harvesting bunds, costs of 143; *see also* integrated watershed management

women 71, 72, 82–3, 86, 101, 139; add value to post-harvest production 83; banking group 89; benefitted from irrigation facilities 106; capacity development for 140; carry water for domestic purposes 89; collectors of shea nuts 52, 144; continued using crop rotation 65; contributions to household food and nutrition security 83; convert 'inefficient' labour time into productive activities 136; differences from men farmers **65**; empowering 52; farmers with less access to land/resources/information 126; farm systems headed by 64, 74; gold 52; granted access to cultivate arable crops 37; have less available labour 70; households headed by **69**, 70, 73, 83, 106, 110, 126; increase the status of 97; labourers **125**; marketing challenges compared to men 83; organized groups 136; registered farmers in Port Harcourt 64; restricted by tradition 15, **26**; skills in processing nuts 56; strengthened in farm-business development 73; traditional practices and skills 140; training on nuts processing 53; yields and gross margin for cassava 68

women, employment of: in aquaculture 78, 82–3, 89; in extension service system 37; opportunities to earn livelihoods from agriculture 136

women, income for: caterpillar as source of 50; generating activities, exclusion from 136; pushed out of traditional sources 15; shea source for 52

women, land access 70; cannot own land 49; farmers with less 126; to harvest shea trees 49; long-term ownership 56; tenure conditions beyond financial capacity of 70; need equal opportunities to own 38

women, participation in group activities 83; community issues 107; decision-making and management 107, 111; meetings and networks 126

women, rural 14; casual job opportunities on farms 89; good management of trees on parklands increases revenues for 56; increase in incomes for 55; not entitled to hold land 37; shea income source for 52

women's livelihoods 16; benefits of shea to 55; importance of shea for 52; values of homegardens to 11

wood 47; fire 1, 50; fuel 2–3, 27, 70, 109; iron 27, *28*; production, revenue from 55; provided by certain tree species 47

woodland: land left fallow regrown into 48; natural 48; natural converted to farm land 47

woody amendments 47; biomass 105; species 49, 54; vegetation 6

yield gaps 11, 14, 143; explanations for 119; potential reduction of 143

yields losses 11, 68; due to parasites 33; through pruning 48

yields 11, *14*, 25, 70, 135, 142–3; advantage 13; decline in 22, **125**; decrease 48; differences in **146**; fish culture systems 81; gains 12; gender gap 71, 72; government policies focus on improving 128; of grain crops 105; greater for alternative practices 144; improved 47, 124; interfered with by shade 55; higher 35, 40, 68,

176 *Index*

yields *continued*
70–1, 122; from intercrop combinations 29; limited potential to optimize 124; literature **99**; multiple crop **125**; new systems that benefit 124; persisting poor 74; potential 11, 119, 143; raising 106; reduced 53; rice 4; sensitive to increasing minimum temperatures 62; shea 52, 55; stagnant 136; *see also* high-yielding
yields, cassava 65, 68, 71; average 62, 66; declines in 73; depend on spacing 66; loss 68; for Nigeria and Thailand 62; trends in multifunctional systems 73–4; for women and men 68, **69**

yields increased/increasing 12, 14, 48, 105, 121, 126, 134, 142, 147; sorghum 53; in Thailand 62
yields, low/lower 16, 53, 62; in women-headed households 126
yields, maize: average *116*, 119, 121; increase in **123**, 126; prioritized optimizing 129; smallholder farmers' 115
yields, millet: increased 53; positive effect on 48

zero tillage 23, 25, **26**, 27, 30–1, **32**, *33–4*, 136